CORE-PLUS MATHEMATICS PROJECT

Course
Part B
1

Contemporary Mathematics in Context
A Unified Approach

Arthur F. Coxford
James T. Fey
Christian R. Hirsch
Harold L. Schoen
Gail Burrill
Eric W. Hart
Ann E. Watkins
with
Mary Jo Messenger
Beth E. Ritsema
Rebecca K. Walker

Glencoe McGraw-Hill

New York, New York   Columbus, Ohio   Chicago, Illinois   Peoria, Illinois   Woodland Hills, California

Glencoe/McGraw-Hill

A Division of The **McGraw·Hill** *Companies*

 This project was supported, in part, by the National Science Foundation.
The opinions expressed are those of the authors and not necessarily those of the Foundation.

Send all inquires to:
Glencoe/McGraw-Hill
8787 Orion Place
Columbus, OH 43240-4027

ISBN: 0-07-827537-7 (Part A)
ISBN: 0-07-827538-5 (Part B)

Contemporary Mathematics in Context
Course 1 Part B Student Edition

7 8 9 10 079/079 10 09 08 07 06

Core-Plus Mathematics Project Development Team

Project Directors
Christian R. Hirsch
Western Michigan University

Arthur F. Coxford
University of Michigan

James T. Fey
University of Maryland

Harold L. Schoen
University of Iowa

Senior Curriculum Developers
Gail Burrill
University of Wisconsin-Madison

Eric W. Hart
Western Michigan University

Ann E. Watkins
California State University, Northridge

Professional Development Coordinator
Beth E. Ritsema
Western Michigan University

Evaluation Coordinator
Steven W. Ziebarth
Western Michigan University

Advisory Board
Diane Briars
Pittsburgh Public Schools

Jeremy Kilpatrick
University of Georgia

Kenneth Ruthven
University of Cambridge

David A. Smith
Duke University

Edna Vasquez
Detroit Renaissance High School

Curriculum Development Consultants
Alverna Champion
Grand Valley State University

Cherie Cornick
Wayne County Alliance for Mathematics and Science

Edgar Edwards
(Formerly) Virginia State Department of Education

Richard Scheaffer
University of Florida

Martha Siegel
Towson University

Edward Silver
University of Michigan

Lee Stiff
North Carolina State University

Technical Coordinator
Wendy Weaver
Western Michigan University

Collaborating Teachers
Emma Ames
Oakland Mills High School, Maryland

Laurie Eyre
Maharishi School, Iowa

Annette Hagelberg
West Delaware High School, Iowa

Cheryl Bach Hedden
Sitka High School, Alaska

Michael J. Link
Central Academy, Iowa

Mary Jo Messenger
Howard County Public Schools, Maryland

Valerie Mills
Ann Arbor Public Schools, Michigan

Marcia Weinhold
Kalamazoo Area Mathematics and Science Center, Michigan

Graduate Assistants
Diane Bean
University of Iowa

Judy Flowers
University of Michigan

Gina Garza-Kling
Western Michigan University

Robin Marcus
University of Maryland

Chris Rasmussen
University of Maryland

Rebecca Walker
Western Michigan University

Production and Support Staff
Lori Bowden
James Laser
Michelle Magers
Cheryl Peters
Angela Reiter
Jennifer Rosenboom
Anna Seif
Kathryn Wright
Teresa Ziebarth
Western Michigan University

Software Developers
Jim Flanders
Colorado Springs, Colorado

Eric Kamischke
Interlochen, Michigan

Core-Plus Mathematics Project Field-Test Sites

Special thanks are extended to these teachers and their students who participated in the testing and evaluation of Course 1.

Ann Arbor Huron High School
Ann Arbor, Michigan
  Kevin Behmer
  Ginger Gajar

Ann Arbor Pioneer High School
Ann Arbor, Michigan
  Jim Brink
  Fay Longhofer
  Brad Miller

Arthur Hill High School
Saginaw, Michigan
  Virginia Abbott
  Felix Bosco
  David Kabobel
  Dick Thomas

Battle Creek Central High School
Battle Creek, Michigan
  Teresa Ballard
  Rose Martin
  Steven Ohs

Bedford High School
Temperance, Michigan
  Ellen Bacon
  David DeGrace
  Linda Martin
  Lynn Parachek

Bloomfield Hills Andover High School
Bloomfield Hills, Michigan
  Jane Briskey
  Homer Hassenzahl
  Cathy King
  Ed Okuniewski
  Mike Shelly

Bloomfield Hills Middle School
Bloomfield Hills, Michigan
  Connie Kelly
  Tim Loula

Brookwood High School
Snellville, Georgia
  Ginny Hanley
  Marie Knox

Caledonia High School
Caledonia, Michigan
  Daryl Bronkema
  Jenny Diekevers
  Thomas Oster
  Gerard Wagner

Centaurus High School
Lafayette, Colorado
  Sally Johnson
  Gail Reichert

Clio High School
Clio, Michigan
  Denny Carlson
  Larry Castonia
  Vern Kamp
  Carol Narrin
  David Sherry

Davison High School
Davison, Michigan
  Evelyn Ailing
  John Bale
  Wayne Desjarlais
  Darlene Tomczak
  Scott Toyzan

Dexter High School
Dexter, Michigan
  Kris Chatas
  Widge Proctor
  Tammy Schirmer

Ellet High School
Akron, Ohio
  Marcia Csipke
  Jim Fillmore
  Scott Slusser

Firestone High School
Akron, Ohio
  Barbara Crucs
  Lori Zupke

Flint Northern High School
Flint, Michigan
  John Moliassa
  Al Wojtowicz

Goodrich High School
Goodrich, Michigan
  Mike Coke
  John Doerr

Grand Blanc High School
Grand Blanc, Michigan
  Charles Carmody
  Nancy Elledge
  Tina Hughes
  Steve Karr
  Mike McLaren

Grass Lake Junior/Senior High School
Grass Lake, Michigan
  Larry Poertner
  Amy Potts

Gull Lake High School
Richland, Michigan
  Virgil Archie
  Darlene Kohrman
  Dorothy Louden

Kalamazoo Central High School
Kalamazoo, Michigan
   Sarah Baca
   Gloria Foster
   Bonnie Frye
   Amy Schwentor

Kelloggsville Public Schools
Wyoming, Michigan
   Jerry Czarnecki
   Steve Ramsey
   John Ritzler

Loy Norrix High School
Kalamazoo, Michigan
   Mary Elliott
   Mike Milka

Midland Valley High School
Langley, South Carolina
   Ron Bell
   Janice Lee

Murray-Wright High School
Detroit, Michigan
   Anna Cannonier
   Jack Sada

North Lamar High School
Paris, Texas
   Tommy Eads
   Barbara Eatherly

Okemos High School
Okemos, Michigan
   Lisa Crites
   Jacqueline Stewart

Portage Northern High School
Portage, Michigan
   Pete Jarrad
   Scott Moore

Prairie High School
Cedar Rapids, Iowa
   Dave LaGrange
   Judy Slezak

San Pasqual High School
Escondido, California
   Damon Blackman
   Ron Peet

Sitka High School
Sitka, Alaska
   Cheryl Bach Hedden
   Dan Langbauer
   Tom Smircich

Sturgis High School
Sturgis, Michigan
   Craig Evans
   Kathy Parkhurst
   Dale Rauh
   Jo Ann Roe
   Kathy Roy

Sweetwater High School
National City, California
   Bill Bokesch
   Joe Pistone

Tecumseh High School
Tecumseh, Michigan
   Jennifer Keffer
   Kathy Kelso
   Elizabeth Lentz
   Carl Novak
   Eric Roberts

Tecumseh Middle School
Tecumseh, Michigan
   Jocelyn Menyhart

Traverse City East Junior High School
Traverse City, Michigan
   Tamie Rosenburg

Traverse City West Junior High School
Traverse City, Michigan
   Ann Post

Vallivue High School
Caldwell, Idaho
   Scott Coulter
   Kathy Harris

West Hills Middle School
Bloomfield Hills, Michigan
   Eileen MacDonald

Ypsilanti High School
Ypsilanti, Michigan
   Keith Kellman
   Mark McClure
   Valerie Mills
   Don Peurach

Overview of Course 1

Part A

Unit  Patterns in Data

Patterns in Data develops student ability to make sense out of real-world data through use of graphical displays and summary statistics.

Topics include distributions of data and their shapes, as displayed in number line plots, histograms, box plots, and stem-and-leaf plots; scatterplots and association; plots over time and trends; measures of center including mean, median, mode, and their properties; measures of variation including percentiles, interquartile range, mean absolute deviation, and their properties; transformations of data.

Lesson 1 *Exploring Data*
Lesson 2 *Shapes and Centers*
Lesson 3 *Variability*
Lesson 4 *Relationships and Trends*
Lesson 5 *Looking Back*

Unit  Patterns of Change

Patterns of Change develops student ability to recognize important patterns of change among variables and to represent those patterns using tables of numerical data, coordinate graphs, verbal descriptions, and symbolic rules.

Topics include coordinate graphs, tables, algebraic formulas (rules), relationships between variables, linear functions, nonlinear functions, and *NOW-NEXT* recurrence relations.

Lesson 1 *Related Variables*
Lesson 2 *What's Next?*
Lesson 3 *Variables and Rules*
Lesson 4 *Linear and Nonlinear Patterns*
Lesson 5 *Looking Back*

Unit  Linear Models

Linear Models develops student confidence and skill in using linear functions to model and solve problems in situations that exhibit constant (or nearly constant) rate of change or slope.

Topics include linear functions, slope of a line, rate of change, intercepts, the distributive property, linear equations (including $y = a + bx$ and *NOW-NEXT* forms), solving linear equations and inequalities, using linear equations to model given data, and determining best-fit lines for scatterplot data.

Lesson 1 *Predicting from Data*
Lesson 2 *Linear Graphs, Tables, and Rules*
Lesson 3 *Linear Equations and Inequalities*
Lesson 4 *Looking Back*

Unit 4 Graph Models

Graph Models develops student ability to use vertex-edge graphs to represent and analyze real-world situations involving relationships among a finite number of elements including scheduling, managing conflicts, and finding efficient routes.

Topics include vertex-edge graph models, optimization, algorithmic problem solving, Euler circuits and paths, matrix representation of graphs, graph coloring, chromatic number, digraphs, and critical path analysis.

Lesson 1 *Careful Planning*
Lesson 2 *Managing Conflicts*
Lesson 3 *Scheduling Large Projects*
Lesson 4 *Looking Back*

Overview of Course 1

Part B

Unit 5 ▶ Patterns in Space and Visualization

Patterns in Space and Visualization develops student visualization skills and an understanding of the properties of space-shapes including symmetry, area, and volume.

Topics include two- and three-dimensional shapes, spatial visualization, perimeter, area, surface area, volume, the Pythagorean Theorem, polygons and their properties, symmetry, isometric transformations (reflections, rotations, translations, glide reflections), one-dimensional strip patterns, tilings of the plane, and the regular (Platonic) solids.

Lesson 1 *The Shape of Things*
Lesson 2 *The Size of Things*
Lesson 3 *The Shapes of Plane Figures*
Lesson 4 *Looking Back*

Unit 6 ▶ Exponential Models

Exponential Models develops student ability to use exponential functions to model and solve problems in situations that exhibit exponential growth or decay.

Topics include exponential growth, exponential functions, fractals, exponential decay, recursion, half-life, compound growth, finding equations to fit exponential patterns in data, and properties of exponents.

Lesson 1 *Exponential Growth*
Lesson 2 *Exponential Decay*
Lesson 3 *Compound Growth*
Lesson 4 *Modeling Exponential Patterns in Data*
Lesson 5 *Looking Back*

Unit 7 ▶ Simulation Models

Simulation Models develops student confidence and skill in using simulation methods—particularly those involving the use of random numbers—to make sense of real-world situations involving chance.

Topics include simulation, frequency tables and their histograms, random-digit tables and random-number generators, independent events, the Law of Large Numbers, and expected number of successes in a series of binomial trials.

Lesson 1 *Simulating Chance Situations*
Lesson 2 *Estimating Expected Values and Probabilities*
Lesson 3 *Simulation and the Law of Large Numbers*
Lesson 4 *Looking Back*

Capstone ▶ Planning a Benefits Carnival

Planning a Benefits Carnival is a thematic, two-week, project-oriented activity that enables students to pull together and apply the important mathematical concepts and methods developed throughout the course.

Contents

Preface

The first three courses in the *Contemporary Mathematics in Context* series provide a common core of broadly useful mathematics for all students. They were developed to prepare students for success in college, in careers, and in daily life in contemporary society. Course 4 formalizes and extends the core program with a focus on the mathematics needed to be successful in college mathematics and statistics courses. The series builds upon the theme of *mathematics as sense-making*. Through investigations of real-life contexts, students develop a rich understanding of important mathematics that makes sense to them and which, in turn, enables them to make sense out of new situations and problems.

Each course in the *Contemporary Mathematics in Context* curriculum shares the following mathematical and instructional features.

- *Unified Content* Each year the curriculum advances students' understanding of mathematics along interwoven strands of algebra and functions, statistics and probability, geometry and trigonometry, and discrete mathematics. These strands are unified by fundamental themes, by common topics, and by mathematical habits of mind or ways of thinking. Developing mathematics each year along multiple strands helps students develop diverse mathematical insights and nurtures their differing strengths and talents.

- *Mathematical Modeling* The curriculum emphasizes mathematical modeling including the processes of data collection, representation, interpretation, prediction, and simulation. The modeling perspective permits students to experience mathematics as a means of making sense of data and problems that arise in diverse contexts within and across cultures.

- *Access and Challenge* The curriculum is designed to make more mathematics accessible to more students while at the same time challenging the most able students. Differences in student performance and interest can be accommodated by the depth and level of abstraction to which core topics are pursued, by the nature and degree of difficulty of applications, and by providing opportunities for student choice on homework tasks and projects.

- *Technology* Numerical, graphics, and programming/link capabilities such as those found on many graphing calculators are assumed and appropriately used throughout the curriculum. This use of technology permits the curriculum and instruction to emphasize multiple representations (verbal, numerical, graphical, and symbolic) and to focus on goals in which mathematical thinking and problem solving are central.

- *Active Learning* Instructional materials promote active learning and teaching centered around collaborative small-group investigations of problem situations followed by teacher-led whole class summarizing activities that lead to analysis, abstraction, and further application of underlying mathematical ideas. Students are actively engaged in exploring, conjecturing, verifying, generalizing, applying, proving, evaluating, and communicating mathematical ideas.

- *Multi-dimensional Assessment* Comprehensive assessment of student understanding and progress through both curriculum-embedded assessment opportunities and supplementary assessment tasks supports instruction and enables monitoring and evaluation of each student's performance in terms of mathematical processes, content, and dispositions.

Unified Mathematics

Contemporary Mathematics in Context is a unified curriculum that replaces the traditional Algebra-Geometry-Advanced Algebra/Trigonometry-Precalculus sequence. Each course features important mathematics drawn from four strands.

The Algebra and Functions strand develops student ability to recognize, represent, and solve problems involving relations among quantitative variables. Central to the development is the use of functions as mathematical models. The key algebraic models in the curriculum are linear, exponential, power, polynomial, logarithmic, rational, and trigonometric functions. Modeling with systems of equations, both linear and nonlinear, is developed. Attention is also given to symbolic reasoning and manipulation.

The primary goal of the Geometry and Trigonometry strand is to develop visual thinking and ability to construct, reason with, interpret, and apply mathematical models of patterns in visual and physical contexts. The focus is on describing patterns with regard to shape, size, and location; representing patterns with drawings, coordinates, or vectors; predicting changes and invariants in shapes; and organizing geometric facts and relationships through deductive reasoning.

The primary role of the Statistics and Probability strand is to develop student ability to analyze data intelligently, to recognize and measure variation, and to understand the patterns that underlie probabilistic situations. The ultimate goal is for students to understand how inferences can be made about a population by looking at a sample from that population. Graphical methods of data analysis, simulations, sampling, and experience with the collection and interpretation of real data are featured.

The Discrete Mathematics strand develops student ability to model and solve problems involving enumeration, sequential change, decision-making in finite settings, and relationships among a finite number of elements. Topics include matrices, vertex-edge graphs, recursion, voting methods, and systematic counting methods (combinatorics). Key themes are discrete mathematical modeling, existence (Is there a solution?), optimization (What is the best solution?), and algorithmic problem-solving (Can you efficiently construct a solution?).

Each of these strands is developed within focused units connected by fundamental ideas such as symmetry, matrices, functions, and data analysis and curve-fitting. The strands also are connected across units by mathematical habits of mind such as visual thinking, recursive thinking, searching for and explaining patterns, making and checking conjectures, reasoning with multiple representations, inventing mathematics, and providing convincing arguments and proofs.

The strands are unified further by the fundamental themes of data, representation, shape, and change. Important mathematical ideas are frequently revisited through this attention to connections within and across strands, enabling students to develop a robust and connected understanding of mathematics.

Active Learning and Teaching

The manner in which students encounter mathematical ideas can contribute significantly to the quality of their learning and the depth of their understanding. *Contemporary Mathematics in Context* units are designed around multi-day lessons centered on big ideas. Lessons are organized around a four-phase cycle of classroom activities, described in the following paragraph—*Launch, Explore, Share and Summarize,* and *On Your Own.* This cycle is designed to engage students in investigating and making sense of problem situations, in constructing important mathematical concepts and methods, in generalizing and proving mathematical relationships, and in communicating both orally and in writing their thinking and the results of their efforts. Most classroom activities are designed to be completed by students working together collaboratively in groups of two to four students.

The launch phase promotes a teacher-led class discussion of a problem situation and of related questions to think about, setting the context for the student work to follow. In the second or explore phase, students investigate more focused problems and questions related to the launch situation. This investigative work is followed by a teacher-led class discussion in which students summarize mathematical ideas developed in their groups, providing an opportunity to construct a shared understanding of important concepts, methods, and approaches. Finally, students are given a task to complete on their own, assessing their initial understanding of the concepts and methods.

Each lesson also includes tasks to engage students in Modeling with, Organizing, Reflecting on, and Extending their mathematical understanding. These MORE tasks are central to the learning goals of each lesson and are intended primarily for individual work outside of class. Selection of tasks for use with a class should be based on student performance and the availability of time and technology. Students can exercise some choice of tasks to pursue, and at times they can be given the opportunity to pose their own problems and questions to investigate.

Multiple Approaches to Assessment

Assessing what students know and are able to do is an integral part of *Contemporary Mathematics in Context*, and there are opportunities for assessment in each phase of the instructional cycle. Initially, as students pursue the investigations that make up the curriculum, the teacher is able to informally assess student understanding of mathematical processes and content and their disposition toward mathematics. At the end of each investigation, the "Checkpoint" and accompanying class discussion provide an opportunity for the teacher to assess levels of understanding that various groups of students have reached as they share and summarize their findings. Finally, the "On Your Own" problems and the tasks in the MORE sets provide further opportunities to assess the level of understanding of each individual student. Quizzes, in-class exams, take-home assessment tasks, and extended projects are included in the teacher resource materials.

Acknowledgments

Development and evaluation of the student text materials, teacher materials, assessments, and calculator software for *Contemporary Mathematics in Context* was funded through a grant from the National Science Foundation to the Core-Plus Mathematics Project (CPMP). We are indebted to Midge Cozzens, Director of the NSF Division of Elementary, Secondary, and Informal Education, and our program officers James Sandefur, Eric Robinson, and John Bradley for their support, understanding, and input.

In addition to the NSF grant, a series of grants from the Dwight D. Eisenhower Higher Education Professional Development Program has helped to provide professional development support for Michigan teachers involved in the testing of each year of the curriculum.

Computing tools are fundamental to the use of *Contemporary Mathematics in Context*. Appreciation is expressed to Texas Instruments and, in particular, Dave Santucci for collaborating with us by providing classroom sets of graphing calculators to field-test schools.

As seen on page iii, CPMP has been a collaborative effort that has drawn on the talents and energies of teams of mathematics educators at several institutions. This diversity of experiences and ideas has been a particular strength of the project. Special thanks is owed to the exceptionally capable support staff at these institutions, particularly at Western Michigan University.

From the outset, our work has been guided by the advice of an international advisory board consisting of Diane Briars (Pittsburgh Public Schools), Jeremy Kilpatrick (University of Georgia), Kenneth Ruthven (University of Cambridge), David A. Smith (Duke University), and Edna Vasquez (Detroit Renaissance High School). Preliminary versions of the curriculum materials also benefited from careful reviews by the following mathematicians and mathematics educators: Alverna Champion (Grand Valley State University), Cherie Cornick (Wayne County Alliance for Mathematics and Science), Edgar Edwards (formerly of the Virginia State Department of Education), Richard Scheaffer (University of Florida), Martha Siegel (Towson University), Edward Silver (University of Michigan), and Lee Stiff (North Carolina State University).

Our gratitude is expressed to the teachers and students in our 41 evaluation sites listed on pages iv and v. Their experiences using pilot- and field-test versions of *Contemporary Mathematics in Context* provided constructive feedback and improvements. We learned a lot together about making mathematics meaningful and accessible to a wide range of students.

A very special thank you is extended to Barbara Janson for her interest and encouragement in publishing a core mathematical sciences curriculum that breaks new ground in terms of content, instructional practices, and student assessment. Finally, we want to acknowledge Eric Karnowski for his thoughtful and careful editorial work and express our appreciation to the editorial staff of Glencoe/McGraw-Hill who contributed to the publication of this program.

To the Student

Contemporary Mathematics in Context may be quite different from other math textbooks you have used. With this text, you will learn mathematics by doing mathematics, not by memorizing "worked out" examples. You will investigate important mathematical ideas and ways of thinking as you try to understand and make sense of realistic situations. Because real-world situations and problems often involve data, shape, change, or chance, you will learn fundamental concepts and methods from several strands of mathematics. In particular, you will develop an understanding of broadly useful ideas from algebra and functions, from statistics and probability, from geometry and trigonometry, and from discrete mathematics. You also will see connections among these strands—how they weave together to form the fabric of mathematics.

Because real-world situations and problems are often open-ended, you will find that there may be more than one correct approach and more than one correct solution. Therefore, you will frequently be asked to explain your ideas. This text will provide you help and practice in reasoning and communicating clearly about mathematics.

Because solving real-world problems often involves teamwork, you often will work collaboratively with a partner or in small groups as you investigate realistic and interesting situations. You will find that two to four students working collaboratively on a problem can often accomplish more than any one of you would working individually. Because technology is commonly used in solving real-world problems, you will use a graphing calculator or computer as a tool to help you understand and make sense of situations and problems you encounter.

You're going to learn a lot of useful mathematics in this course—and it's going to make sense to you. You're going to learn a lot about working cooperatively and communicating with others as well. You're also going to learn how to use technological tools intelligently and effectively. Finally, you'll have plenty of opportunities to be creative and inventive. Enjoy.

Patterns in Space and Visualization

The Shape of Things

The world, the space in which you live, is three-dimensional. Buildings, plants, animals, toys, tools, even molecules are three-dimensional; they are **space-shapes**. Space-shapes develop in nature and are built, observed, and used in all countries and cultures. They can be beautiful, unusual, large, or small. If a space-shape is to serve a purpose, it may need to have special characteristics. For example, it may need to bounce or resist strong wind forces. It also may need to support great weight, use space efficiently, or be hollow or filled.

Think About This Situation

Look at the photo of Biosphere 2 shown above and at objects around your classroom. Identify two or three space-shapes.

a List some characteristics of the space-shapes you see.

b How are two-dimensional shapes (*plane-shapes*) used to make these space-shapes?

c Choose one of the space-shapes in your classroom.

■ Draw a sketch of it.

■ Explain how you might use numbers to describe its size.

INVESTIGATION 1 Designing and Testing Columns

Space-shapes come in all shapes and sizes and have many uses. For example, many ancient cultures used filled or solid space-shapes to make columns in their buildings. The Greek Parthenon shown here is made of marble. Thus, the columns had to be designed to support great weights.

The columns the Greeks used were solid marble. You can model the Greek columns with *shells* rather than solids. That is, your columns will be hollow like tin cans (with no top or bottom) rather than filled like hockey pucks. In this investigation, you will seek an answer to the question:

"What type of column supports the most weight?"

1. As a group, brainstorm about a possible answer to this question. What column characteristics do you think support your choice?

2. Now, work in pairs to complete the experiment described below.

   a. Select four sheets of 8.5×11-inch paper such as typing paper, copy machine paper, or computer printer paper.

   b. Make four columns, each 8.5 inches high. The bases of the columns should have the following shapes:

   - triangular with all sides equal;
   - square;
   - eight-sided with all sides equal;
   - circular.

   For consistency, leave a half-inch overlap and tape the columns closed. Be sure to tape near each end.

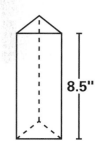

8.5"

   c. Set up the following weight-supporting situation to collect data.

   - Choose a level surface.
   - Place a small rectangle of cardboard (about 6×8 inches) on top of a column.
   - Choose a sequence of objects to be placed on the cardboard platform. (**Note:** Be sure to use the same sequence for each test.)
   - Carefully add objects until the column collapses. Measure and record the maximum weight supported.
   - Organize and display your data in a table.

3. Make a graph of your data with *number of sides* on the horizontal axis and *maximum weight supported* on the vertical axis.

    a. Where along the horizontal axis did you put "circular column"? Why?

    b. What appears to happen to the maximum weight supported as the number of sides of the column increases?

    c. Use your table or graph to estimate the weight-supporting capacity of a 6-sided column. To check your estimate, make a 6-sided column and find the maximum weight it supports.

Checkpoint

In this investigation, you explored the weight-bearing capacity of differently-designed columns.

a Why do you think the ancient Greeks chose to use cylindrical columns?

b What are some other questions about column design that seem important and which you could answer by experimentation?

Be prepared to share your group's thinking and questions with the class.

On Your Own

Latoya investigated what happened to the amount of weight supported when she increased the number of columns underneath the cardboard platform. She kept the shape and area of the base the same for all columns in each experiment. Her data for the three experiments are summarized in the table below.

Triangular Columns		Square Columns		Circular Columns	
Number of Columns	Weight (kg) Supported	Number of Columns	Weight (kg) Supported	Number of Columns	Weight (kg) Supported
1	1.7	1	2.1	1	3.3
2	3.5	2	4.1	2	6.4
3	5.1	3	6.4	3	10.0
4	7.0	4	8.5	4	13.2
5	8.3	5	10.4	5	16.6

a. Make a scatterplot of Latoya's data. Use a different symbol for the data from each of the three experiments.

b. What do you think is true about the relationship between *number of columns* and *weight supported*?

c. How are the patterns of change in *weight supported* as the *number of columns* increases similar for the different types of columns? How are they different?

d. For each type of column, estimate the number of columns needed to support a weight of 20 kg. Explain your method.

INVESTIGATION 2 Recognizing and Constructing Space-Shapes

The columns studied in Investigation 1 are examples of space-shapes. Most everyday space-shapes are designed with special characteristics in mind.

1. As a class, examine the space-shapes depicted below.

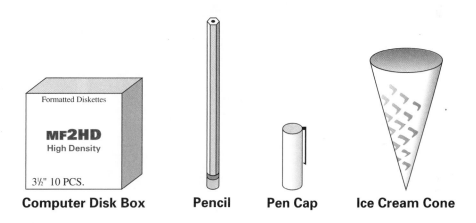

| Computer Disk Box | Pencil | Pen Cap | Ice Cream Cone |

a. Which shapes above have common characteristics? What are those characteristics?

b. In what ways might the word "parallel" be used to describe characteristics of some of the space-shapes?

2. On the following page are pictures of structures built through the centuries by various cultures. Describe the different space-shapes you see in these photographs. Name those that you can. Are some forms of space-shapes more common in some cultures than in others?

a.

Igloo in the Arctic

b.

Pyramid at Chichen Itza, Mexico

c.

Native American Tepees

d.

Stave Church, Norway

e.

Himeji Castle, Japan

f.

Taj Mahal, India

3. Two important classifications of space-shapes are **prisms** and **pyramids**. In your group, study the examples and non-examples below.

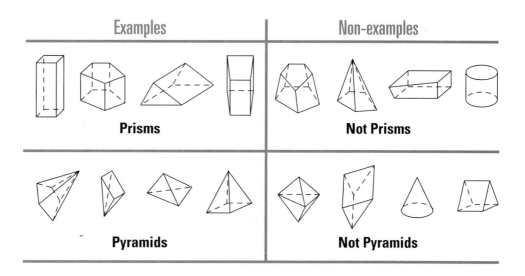

Examples	Non-examples
Prisms	**Not Prisms**
Pyramids	**Not Pyramids**

  a. Which of the space-shapes below are prisms? Explain your reasoning.

  b. Which of the space-shapes below are pyramids? Explain your reasoning.

  c. Explain why the remaining space-shapes are neither pyramids nor prisms.

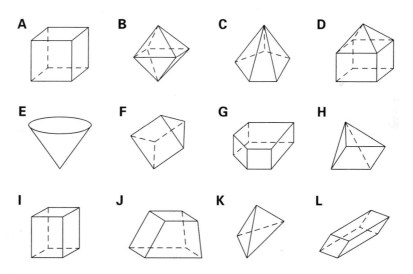

4. Now try to describe the distinguishing features of pyramids and prisms.

  a. What are the characteristics of a space-shape that make it a pyramid?

  b. What are the characteristics of a space-shape that make it a prism?

5. **Cones** and **cylinders** are two other common space-shapes. How are they similar to pyramids and prisms? How are they different?

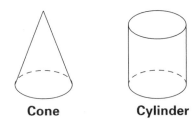

Cone   **Cylinder**

6. Develop a method for naming types of prisms and types of pyramids. Could you use the shape of the base in your naming procedure? Try it.

   a. Use your method to name a cube as a type of prism.

   b. Cardboard boxes are usually prisms. What would you name such a prism?

Checkpoint

In this investigation, you discovered identifying characteristics of prisms and pyramids.

a How are pyramids similar to prisms? How are they different?

b Sketch a prism in which the bases are five-sided. Name the prism.

c What space-shape best describes a Native American tepee?

Be prepared to share your descriptions and sketch with the class.

▶ On Your Own

An "A-Frame" is a style of architecture sometimes used in building houses. What space-shape is basic to the "A-Frame" construction?

Space-shapes can be modeled in three different ways. One way is as a *solid object* such as a brick, a cake, or a sugar cube. Another is as a *shell* such as a paper bag, a house, or a water pipe. The third way is as a *skeleton*, which includes only the edges like a jungle gym. A solid model can be made of material such as clay, wood, or plastic foam. A shell model can be made of paper or cardboard and tape, for example. A skeleton model can be made with things like toothpicks and clay or straws and pipe cleaners.

7. Use a piece of modeling clay or Play-Doh to make a cube.

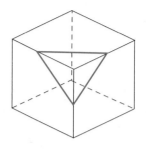

   a. How many flat surfaces or *faces* does it have?

   b. Imagine slicing a corner off. Now how many faces are there? What would you see as the shape of the new face on the cube? This is a **plane slice** since the new face is flat. Slice your cube to check your prediction.

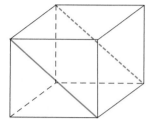

   c. Imagine a line joining two opposite corners of a face. Make a plane slice all the way through to the corresponding line on the opposite face. What is the shape of the new face formed? Check by slicing your cube. What do you notice about the two pieces of the cube?

   d. Think of two other ways of slicing your cube (remade, of course). Predict and sketch the shape of the new face. Do the slicing to check your predictions.

8. Get a collection of straws and pipe cleaners from your teacher. Your group will need about 75 of the 10-cm straws, 27 of the 12-cm straws, and 6 of the 16-cm straws. Cut the pipe cleaners into 5- or 6-centimeter length pieces and bend them in half. These pieces will be used to connect two straw edges at a vertex, as shown here. Carefully cut the straws into the lengths given in Part a.

   a. Make the following models from straws and pipe cleaners. Divide the work among the group members. Each student should build at least one prism and one pyramid.

   - cube: 10-cm edges
   - triangular prism: 10-cm edges on bases, 12-cm height
   - square prism: 10-cm edges on bases, 12-cm height
   - pentagonal prism: 10-cm edges on bases, 12-cm height
   - hexagonal prism: 10-cm edges on bases, 12-cm height
   - triangular pyramid: 10-cm edges on bases, other edges 10 cm
   - square pyramid: 10-cm edges on bases, other edges 12 cm
   - pentagonal pyramid: 10-cm edges on bases, other edges 12 cm
   - hexagonal pyramid: 10-cm edges on bases, other edges 16 cm

b. Imagine slicing your triangular prism with a single plane slice. What is the shape of the new face produced by a single plane slice? Check by slicing a clay model if you have any doubts.

c. Think about the shapes of faces that are created when the triangular prism is sliced. Identify as many *different* shapes of these faces as you can. Sketch the shape of each new face. Check each with clay models if it is helpful.

d. Now imagine slicing each of the other prisms with planes. What new face shapes do you get with a single plane slice? Sketch each, and check with clay models if group members are not all in agreement.

e. Do any of your plane slices produce two halves that are identical space-shapes? If so, describe the slice location(s).

Some planes may slice a space-shape into two identical mirror-image halves. When this is possible, the space-shape is said to have the property of **reflection symmetry**. The plane is called a **symmetry plane** for the space-shape. For example, when you want to share a piece of cake fairly, you cut it into two identical mirror-image halves.

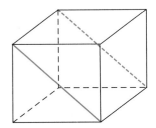

9. Which of the space-shapes constructed in Activity 8 have reflection symmetry? Find and describe all the symmetry planes for each shape. Compare your results with those of other groups.

Checkpoint

Visualizing planes slicing a space-shape can help in identifying properties of the shape.

a Imagine a prism and a plane. Describe several ways the plane could slice the prism without intersecting either base.

b Think about how the symmetry planes for prisms and pyramids are related to the bases. Explain how the symmetry planes differ for the two kinds of space-shapes.

Be prepared to share your ideas with the entire class.

▶ **On Your Own**

Imagine or construct a rectangular prism. Determine the number of symmetry planes for the shape. Describe their locations.

MORE
Modeling • Organizing • Reflecting • Extending

Modeling

1. What do you think is true about the relationship between the height of a square column and the weight it can support?

   a. With a partner, investigate the weight-bearing capability of square columns of different heights. Vary the column heights, but keep the shape and area of the base the same.

   b. Organize your data in a table and display them in a graph.

   c. What appears to be true about the relationship between the height of a column and the weight it can support?

2. Make a conjecture about the relationship between the circumference of a circular column and the weight it can support.

   a. With a partner, conduct an experiment to test your conjecture about the weight-bearing capability of circular columns from this new perspective. Use columns of the same height, but with different circumferences.

   b. Organize your data in a table and display them in a graph.

   c. What appears to be true about the relationship between the circumference of a column and the weight it can support? Why do you think this happens?

3. Space-shapes form the basis of atomic structures as well as of common structures for work, living, and play. Often a single space-shape is not used, but rather a combination.

   a. Study this photograph of a tower in Europe. What space-shapes appear to be used in this tower?

   b. Scientists use space-shapes to model molecules of compounds. Shown at the left is a model of a methane molecule. Describe the space-shape whose skeleton would be formed by joining the four hydrogen atoms that are equally-spaced around the central carbon atom. Describe all possible planes of symmetry.

Big Ben clock tower in London, England

c. The square pyramids at Giza are pictured here. Describe one so that some-one who had never seen it could visualize it in his or her mind. Describe all the possible planes of symmetry of one of the pyramids.

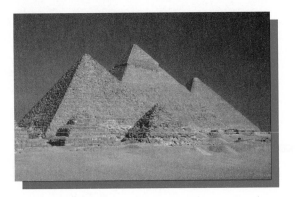

4. Examine the labeling of the faces of a die (one of several dice).

a. How are the dots (called *pips*) on opposite faces related?

b. Use cubes made of wood or clay to help you determine how many different ways the pips can be put on the faces so that opposite faces add to seven.

c. Place your models on your desk so that the face with one pip is on the top and the face with two pips is toward you. How are the models related to each other?

Organizing

1. Refer to the model of a cube in Activity 7 of Investigation 2.

a. How many faces, edges, and vertices does it have?

b. Slice a corner off (as shown), making a small triangular face. Repeat at each cor-ner so that the slices do not overlap. Make a table showing the number of faces, edges, and vertices of the modified cube after each "corner slice."

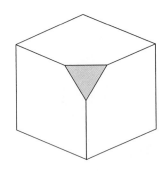

c. Using *NOW* and *NEXT*, write a rule des-cribing the pattern of change in the number of faces after a slice. Write similar *NOW-NEXT* rules for the number of edges and for the number of vertices after each slice.

d. How many faces, edges, and vertices does the new solid have when all the corners are sliced off?

2. A mirror acts like a symmetry plane for you and your "mirror" image.

   a. If you walk toward a mirror, what appears to happen to your image?

   b. If your nose is one meter from a mirror, how far does your "image" nose appear to be from the mirror?

   c. Imagine tying your index finger to its mirror image with a taut rubber band. How would the rubber band (a segment) be related to the mirror (a plane)? Would this relationship change as you moved your finger? Why or why not?

3. The figure at the right shows a space-shape and one of its symmetry planes. Points *A* and *B* are on the space-shape and are symmetrically placed with respect to the plane.

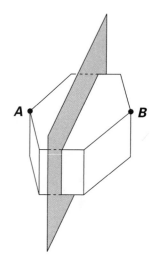

   a. If you connect *A* and *B* with a segment, would it intersect the symmetry plane?

   b. How is $\overline{AB}$ (segment *AB*) related to the symmetry plane?

   c. Compare the distances from *A* and from *B* to the symmetry plane.

   d. How can the word "perpendicular" be used in discussing a symmetry plane and any two symmetrically placed points such as *A* and *B*?

4. The first three elements of a sequence of staircases made from cubes glued together are shown here.

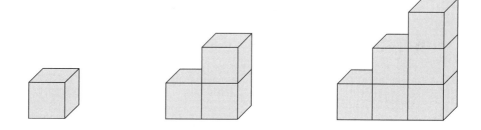

   a. Sketch the 4th and 5th staircases.

   b. If the 4th staircase has 10 blocks, how can you determine the number of blocks in the 5th staircase without counting?

   c. If the staircase in the *n*th position has *B* blocks in it, how many blocks are in the next staircase?

   d. Describe any symmetry planes for the staircases.

e. Imagine taking a staircase and fitting a copy of it on top to form a rectangular prism. In this way, form rectangular prisms from pairs of the first, second, fourth, and fifth staircases. Answer the following questions.

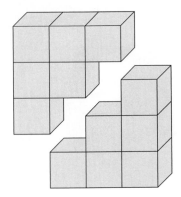

- How many blocks are in each prism?
- How is the number of blocks in each prism related to the number of blocks in each original staircase?
- How can you use the number of blocks on the bottom row of a staircase to predict the number of blocks in the prism?
- How can you use the number of blocks on the bottom row of a staircase to predict the total number of blocks in the staircase?

f. Write a formula for predicting the number of blocks in a staircase with n blocks on the bottom row.

- Use your formula to predict the total number of blocks in a staircase with 10 blocks on the bottom row.
- Use your result in Part c to help check your answer.
- How many blocks are on the bottom row of a staircase made up of 276 blocks?

g. What happens when you fit a staircase to the one immediately following it in the sequence of staircases?

Reflecting

1. Pyramids were built by the Egyptians as tombs for their rulers. What other cultures built structures shaped as pyramids? What purposes did they serve?

2. Think about the similarities and differences between prisms and cylinders as you complete this task.

a. Explain why every prism has at least one symmetry plane.

- Draw a prism, then sketch in a symmetry plane whose description could involve the idea of *parallel*.
- The symmetry plane you drew for Part a could also be described using the idea of *perpendicular*. To what is the symmetry plane perpendicular?

b. Describe the symmetry planes for a prism with an equilateral, equiangular pentagonal base.

c. Describe the symmetry planes for a cylinder.

3. Cedar posts are circular columns often used for building fences.

   a. How could you pack cedar posts for shipment? Sketch your arrangement (the end view).

   b. Are other post-packing arrangements possible? Illustrate those you find with sketches.

   c. In your experience, what consumer goods have you seen packaged in the manner of Part a? In the manner of Part b?

4. Reflection symmetry (sometimes called *bilateral symmetry*) is often found in nature. What examples of bilateral symmetry in nature have you seen in your science classes?

Extending

1. A cube is a square prism with all faces and bases congruent. For each figure below, describe how a plane and a cube could intersect so that the intersection is the figure given. If the figure is not possible, explain your reasoning.

   a. a point   **b.** a segment

   c. a triangle   **d.** an equilateral triangle

   e. a square   **f.** a rectangle

   g. a five-sided shape   **h.** a six-sided shape

2. Recall your column-building experiment in Investigation 1. Can you improve the weight-bearing capability of the poorest-performing column to equal the capability of the best-performing column? Choose one of the two methods below. Describe a procedure you could use to see if the method will improve the column's performance. Conduct the experiment and write a summary of your findings.

   a. Increase the thickness of the walls of a column.

   b. Increase the number of columns used to support weight.

3. The prisms you examined in Investigation 2 are right prisms because the edges connecting the bases are perpendicular to the edges of the bases. Use your skeleton models to help you visualize prisms where these edges are *not perpendicular* to the edges of the bases. These prisms are called **oblique**.

   a. What shapes are the faces of an oblique prism?

   b. Look up the definition of a prism in the dictionary. Is a particular kind of prism defined? If so, what kind?

   c. Do oblique prisms have symmetry planes?

   d. Given an oblique prism with five or fewer faces, imagine its intersection with a plane. What is the shape of the intersection? Find as many such shapes as you can. Name each of them.

4. Suppose each face of a cube is painted one of six different colors. How many cubes with different coloring patterns are possible?

5. Make six cubes with different coloring patterns, as described in Extending Task 4. Is it possible to join them together in a row so that all faces along the row are the same color, the colors of the touching faces match, and the colors of the two end faces match? If so, display your solution.

INVESTIGATION 3 Visualizing and Sketching Space-Shapes

Models of space-shapes are valuable for several reasons. For an architect, a scale model of a building gives the client a visual impression of the finished product. When the characteristics of a space-shape need to be understood, models allow those characteristics to be more easily visualized and verified.

Model of Biosphere 2

However, it is not always practical to construct models of space-shapes. For example, you cannot fax a scale model of an off-shore oil rig to a Saudi engineer. Rather, the space-shape needs to be represented in two dimensions and still convey the important information about the shape. In this investigation, you will explore various ways of drawing and sketching space-shapes.

One way to depict space-shapes is to draw the shape from various views. A method commonly used by architects is to draw **face-views**. For the house at the right, a *top view*, a *front view*, and a *right-side view* are shown below. Together, these views display the width, depth, and height of the building. (You'll notice the top view is different from the other two. Frequently, floor plans such as this are used instead of an exterior top view.)

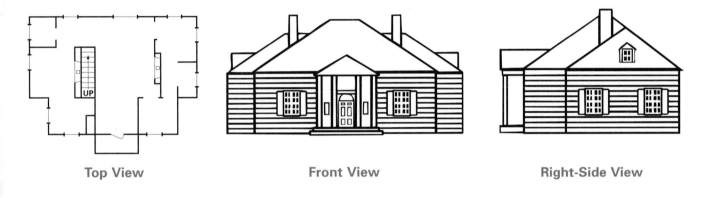

Top View   **Front View**   **Right-Side View**

1. Below are three face-views of a simple model of a hotel made from cubes.

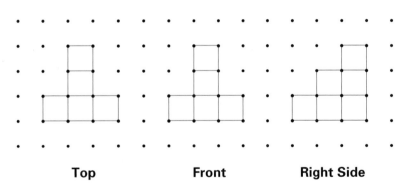

Top   **Front**   **Right Side**

a. How many cubes are in the model?

b. Use blocks or sugar cubes to make a model of this hotel. Build your model on a sheet of paper or poster board that can be rotated.

c. Could you make the model using information from only two of these views? Explain your reasoning.

2. Shown here is a *top-front-left corner view* of a model hotel drawn on *isometric* dot paper.

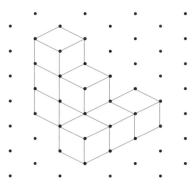

a. Try to rotate your model from Activity 1 so that it appears as in this **isometric drawing**.

b. Describe as completely as possible the vantage point from which the model is being viewed in this drawing.

c. On your own, use isometric dot paper to draw a top-front-right corner view of your model. Compare your drawing with those of others in your group.

d. Use dot paper to draw an isometric view of a model hotel that has top, front, and right-side views as shown below.

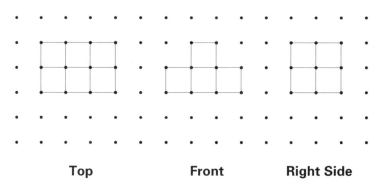

Top   **Front**   **Right Side**

e. Compare your drawing with those of other group members.

 ■ Does each drawing accurately depict the hotel? If not, work together to modify any inaccurate drawings.

 ■ Describe as well as you can the vantage point from which the hotel is viewed in each drawing.

f. Draw top, front, and right-side views for the model hotel below. Assume there are no hidden cubes.

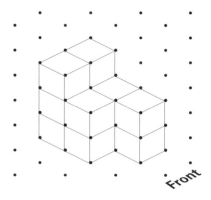

3. Shown below is another way to represent a space-shape such as a box of computer disks. The sketch on the right provides a top-front-right corner view of the box as a geometer would draw it. The sketch preserves a sense of depth even though it is not drawn in true perspective. **Hidden lines**, such as the three back edges of the box, are shown as *dashed lines*.

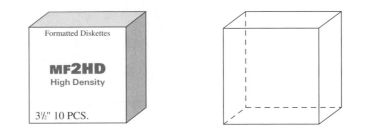

a. What space-shape is a box of computer disks?

b. How are the shapes shown in the three face-views below related to the actual faces of the box?

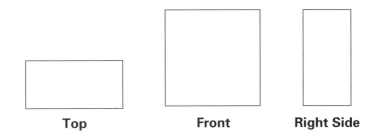

c. Now examine more carefully the sketch of the box from a top-front-right corner view.

- What appear to be the shapes of the faces as shown in the drawing? What are the true shapes of the faces on the box itself?

- In this type of sketch, how do you know how long to draw each edge?

- What edges are parallel in the real object? Should the corresponding edges in the sketch be drawn parallel? Explain your reasoning.

4. For this activity, use the prisms you constructed from straws and pipe cleaners for Activity 8 on page 333.

a. Draw top, front, and right-side views of each of the prism models.

b. Now sketch each of your prism models from a top-front-right corner view. Show the hidden edges.

c. Each member of your group should select a different prism. Sketch your prism from a top-front-left corner view. Check each others' sketches and suggest improvements as necessary.

5. Compare an isometric drawing of a cube to the sketch of a cube you prepared in Activity 4 Part b. How are they similar and how are they different?

6. For this activity, refer to your pyramid models made in Activity 8, page 333.

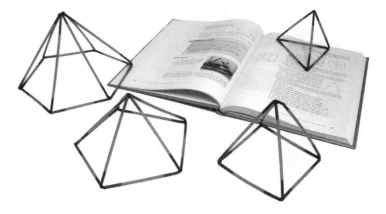

  a. Sketch each of these pyramid models from a view in front and above the base. Show all hidden edges.

  b. How does drawing hidden edges help you visualize the entire space-shape?

7. Compare representations of space-shapes by drawing face-views, by making isometric drawings, and by making sketches from particular viewing points. What are the advantages and drawbacks of each method?

Checkpoint

A space-shape can be represented in two dimensions by a face-views drawing, an isometric drawing, or a geometric sketch.

  ⓐ Make a model of a space-shape that is two square pyramids sharing a common base. Represent the two-piece shape using the method you think is best.

  ⓑ Explain your reasons for using the method you did.

  Be prepared to share your drawing and discuss reasons for your choice of method.

▶ On Your Own

Sketch the space-shape formed when a pentagonal pyramid is placed on top of a pentagonal prism. Describe a possible real-world application of a shape with this design.

INVESTIGATION 4 Rigidity of Space-Shapes

Buildings, bridges, and other outdoor structures must withstand great forces from the environment. While there are many types of space-shapes, only certain kinds of shapes are used to make structures which do not collapse under pressure.

1. As a class, examine the frameworks used to design the Statue of Liberty and the Eiffel Tower, as shown.

   a. What simple plane shape is fundamental to these frames?

   b. Why do you think Gustave Eiffel used this shape as the basis for his design of both structures?

2. Working in groups, examine the *rigidity* of your models from Activity 8 of Investigation 2, page 333.

   a. Which of these space-shapes are **rigid**? That is, which of the shapes will not change form when a force is applied to any part of it? Consider cases where a base of the shape rests on a plane surface and where the shape does not rest on a surface.

   b. Do the rigid space-shapes have anything in common? Explain.

3. Add reinforcing straws to your model of a triangular prism to make it rigid.

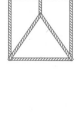

   a. How many reinforcements did you use? Describe where you placed them and why you placed them there.

   b. Could you have placed the reinforcements in different positions and still made the triangular prism rigid? Explain and illustrate.

4. Add reinforcements to your model of a cube so that it becomes a rigid structure.

   a. Note the number of reinforcing straws that you used and describe the position of each straw.

   b. Find a different way to reinforce the cube so that it becomes a rigid structure. Describe the pattern of reinforcing straws.

   c. Of the methods you used to reinforce the cube, which could be used to make a rectangular building stand rigidly?

5. Now consider your straw model of a pentagonal prism.

   a. Predict the minimum number of reinforcing straws needed to make it rigid.

   b. Make the prism rigid and compare the number of reinforcing straws needed with your prediction. Can you find a way to make the prism rigid with fewer reinforcements?

Checkpoint

Rigidity is often an important consideration in designing a space-shape.

a What is the simplest rigid space-shape?

b What reinforcement patterns are used to make a space-shape rigid?

Be prepared to share your group's findings with the class.

On Your Own

How many reinforcing straws would be needed to make a straw model of a hexagonal prism rigid? Where would you place the reinforcements? Draw a sketch of the prism with reinforcements.

Modeling • Organizing • Reflecting • Extending

Modeling

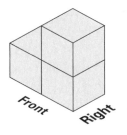

1. Designers of structures like motels can test their designs by using identical cubes to represent the rooms. They can use the cubes to try various arrangements of rooms. Suppose you have a three-room motel and that cubes must join face-to-face. A sample model is shown at the right.

   a. How many different three-room motels can you construct?

   b. Make top, front, and right-side view sketches of each possible motel.

   c. Make isometric drawings of the same motels. What vantage point did you use for the drawings?

   d. Which motel would be the least costly to construct? Why?

2. Increase the number of rooms in your motel from three to four.

   a. Find all the four-room motels you can build if rooms must connect face-to-face. How many did you find?

   b. Make face-view sketches or isometric drawings of each motel.

   c. How many of your motels form an L-shape?

   d. Land in and around cities is very expensive. Which of your motels would require that the least amount of land be purchased? The greatest amount? Explain your reasoning.

3. Study this drawing of a cube motel.

   a. How many cubes are there in the model?

   b. Draw the top, front, and right-side views of this shape.

   c. Suppose this model is half of a double towers structure. The towers are symmetrically built about the plane of the right side. Make a sketch or an isometric drawing of the complete structure.

   d. Provide top, front, and right-side view sketches of the completed building.

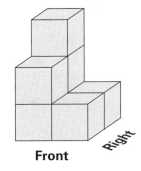

4. Below are three views of a cube model of a hotel.

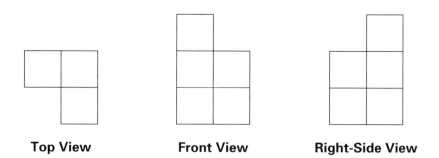

Top View   **Front View**   **Right-Side View**

  a. Construct cube models which match these views.

  b. Can you construct more than one model? If so, explain how they differ.

  c. Make a three-dimensional drawing of your model. Choose a vantage point that shows clearly all the characteristics of your model.

5. Both portability and rigidity are design features of a folding "director's chair."

  a. How are these features designed into the chair shown at the right?

  b. Identify at least two other items which must remain rigid when "unfolded" and analyze their designs.

Organizing

1. In Investigation 3, you learned how to represent three-dimensional space-shapes in two dimensions. Another way to do this with a rectangular prism is illustrated below.

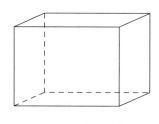   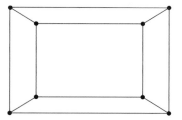

Rectangular Prism   **Vertex-Edge Graph**

a. The figure on the right at the bottom of page 348 is a representation of a rectangular prism as a vertex-edge graph with no edge crossings. You can think of this graph resulting from "compressing" a rectangular prism with elastic edges down into two dimensions. The graph does not look much like the three-dimensional prism, but it shares many of the prism's properties. Name as many shared properties as you can.

b. Use the idea of *compressing* illustrated in Part a to draw vertex-edge graphs representing the following polyhedra. After compressing shape iii, non-intersecting edges in the shape should not intersect in the graph. You will need to stretch the edges that go to one vertex to be sure this does not happen.

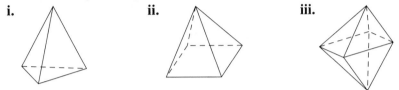

i.   **ii.**   **iii.**

2. In this task, you will re-examine the straw space-shapes you constructed. Count the edges, vertices (points where two or more edges meet), and faces of each space-shape. Organize your data in a table.

a. Look for a pattern relating the number of faces, vertices, and edges for each shape. Describe any patterns you see.

b. In Organizing Task 2 (page 262) of the "Graph Models" unit, you may have discovered that the numbers of vertices V and regions R formed by edges E of certain graphs are related by the formula $V + R - E = 2$. Compare this pattern to your pattern for the shapes in Part a. Write a similar equation relating the numbers of vertices V, faces F, and edges E of each of your space-shapes.

c. Write two equations that are equivalent to the equation $V + R - E = 2$. Which form of the equation would be easiest to remember? Why?

3. There are interesting and useful connections between pairs of special space-shapes.

a. Examine the triangular pyramid constructed from six equal-length straws. It is called a **regular tetrahedron**. Imagine the centers of each face.

- How many such centers are there?

- Imagine connecting the centers with segments. How many such segments are there?

- Visualize the space-skeleton formed by the segments. What are the shapes of its faces?

- What space-shape would be formed by the segments?

b. Examine visually the straw model of a cube. Imagine the centers of each face.

- How many such centers are there?

- Imagine connecting, with segments, each center to the centers on the four **adjacent faces**. Adjacent faces are faces that have a common edge. How many such segments are there?

- Visualize the space-skeleton formed by the segments. What are the shapes of its faces? How many faces are there?

c. Make a straw model of the space-skeleton in Part b. This shape is called a **regular octahedron**.

4. Bridge trusses are examples of rigid space-shapes. Shown here is a Parker truss in Elizabeth, New Jersey.

In addition to the Parker truss, there are several other types of trusses that vary in cost, strength, and other factors. Three of them are illustrated below.

i.   **ii.**   **iii.**

Pratt Truss   **Warren Truss**   **Howe Truss**

a. Determine if it is possible to make a model of each using a single piece of wire. If possible, draw sketches showing the manner in which you would bend the wire.

b. Explain your results in terms of vertex-edge graphs and Euler paths.

5. Here is a quick way to make a tetrahedron from a sealed letter envelope.

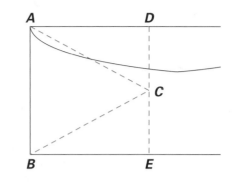

- Find point *C* so that segments *AB*, *AC*, and *BC*, are the same length. (*Hint:* Try folding, measuring, or using a compass.)
- Cut along $\overline{DE}$ parallel to $\overline{AB}$.
- Fold back and forth along $\overline{AC}$ and $\overline{BC}$.
- Open the envelope at the cut, and then pinch the envelope so that points *D* and *E* are together. Tape along the cut edge.

a. Label the faces of the tetrahedron 1, 2, 3, and 4 to make a tetrahedral die. Toss the die 100 times. Record the number of times each face lands down.

b. Make a histogram of your data.

c. Describe the shape of the histogram.

6. Obtain or make a tetrahedral die as described in Organizing Task 5. Gather data on the following experiment. Use a table to organize your work.

- Toss the die 10 times; record the number of times the face labeled 3 lands down. Determine the ratio: number of 3s ÷ total tosses (10).
- Toss 10 more times and determine the ratio: number of 3s (in both sets of tosses) ÷ total tosses (20).
- Repeat until you have 100 tosses.

a. Create a graph with "number of tosses" on the horizontal axis and "ratio of 3s to number of tosses" on the vertical axis.

b. What does the graph tell you?

Reflecting

1. In this lesson, as well as in previous units, you have engaged in important kinds of mathematical thinking. From time to time, it is helpful to step back and think about the kinds of thinking that are broadly useful in doing mathematics. Look back over the four investigations in this lesson and consider some of the mathematical thinking you have done. Describe an example where you did each of the following:

a. experiment;

b. search for patterns;

     c. formulate or find a mathematical model such as a function rule;

     d. visualize;

     e. make and check conjectures;

     f. make connections

        ■ between mathematics and the real world;

        ■ between mathematical strands (between geometry and algebra, geometry and statistics, or geometry and graph theory).

2. Which of the methods of drawing space-shapes is most difficult for you? Why?

3. Which drawing method provides the most complete information regarding a space-shape? Explain your reasoning.

4. A particular space-shape is symmetrical about a plane perpendicular to the front of the shape. What can you conclude about the front view of the shape? Why?

5. Buildings in areas that are subject to the stresses of earthquakes must have "flex" built into them. How can a space-shape have both rigidity and flex?

Extending

1. Models of solids can be made by folding a pattern drawn on paper and taping or gluing the edges. Here is such a pattern for a square pyramid. Find and sketch two other patterns that would fold into a square pyramid. How many straight cuts are needed in order to cut out your patterns?

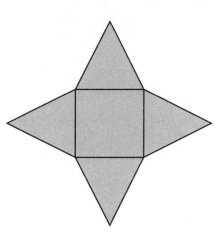

2. Try to imagine a space-shape that has a rectangle as its front view, a triangle as its side view, and a circular top view. Is such a space-shape possible? If so, sketch it. If not, explain why it is impossible.

3. Use your results from Modeling Tasks 1 and 2 to identify all the three- and four-room motels with right-angle bends in them. If possible, make a model of each motel using cubes taped or glued together. For some of the motels, there is another that is actually the same space-shape. Only include one of these motels.

a. How many motels with right-angle bends are there?

b. How many individual cubes are needed to make the whole collection of these shapes?

c. Which of the shapes have reflection symmetry? For shapes that are symmetric, write the number of planes of symmetry.

d. Are any two of your shapes related by symmetry? If so, identify them.

4. One of the strongest types of structures is a dome. However, it is very difficult to build a dome without the weight of the building material being too great. To lighten the weight, dome-like space frames are used to approximate the shape of a true dome. These dome-like space frames are generally referred to as *geodesic domes*. The work of R. Buckminster Fuller greatly influenced the popularity of the dome. A picture of Fuller's dome at the Montreal Exposition in 1967 is shown below.

R. Buckminster Fuller, left, discusses geodesic design with an artist.

Montreal Exposition, 1967

a. Look up the meaning of the roots of the word "geodesic." Explain why you think Fuller used this term to describe the dome.

b. What are the fundamental units of the Expo dome? What other polygons do you notice in this dome?

c. Conduct research on other geodesic domes such as the Epcot Center at Disney World in Orlando, Florida. What are the fundamental units of those domes? What other polygons do you notice in the domes?

5. Some space-shapes can be made by weaving strips of paper together. Obtain a copy of the following patterns.

a. Cut along segment *ST*, making two strips of five equilateral triangles. Letter the corners as shown. Fold along the sides of the triangles. Try to weave these strips into a tetrahedron. If you need assistance, use the instructions below the figures. "$A_2 \rightarrow A_1$" means face A_2 goes under face A_1 with the letters on top of each other.

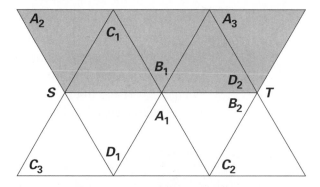

Weaving Procedure: $A_2 \rightarrow A_1$; $C_2 \rightarrow C_1$; $B_2 \rightarrow B_1$; $A_3 \rightarrow A_1$ and A_2; slide C_3 through the open slot so $C_3 \rightarrow C_1$ and C_2, and $D_2 \rightarrow D_1$.

b. You can also weave a cube. Cut out the 3 strips from a copy of this pattern and follow the weaving procedure below the figure. "$A_2 \rightarrow A_1$" means face A_2 goes under face A_1 with the letters on top of each other. "$D_2 (D_3)$" means face D_2 with face D_3 under it.

		D_1	A_2	B_1	
E_3	F_2			A_1	E_2 F_3
	F_1	C_2			
	C_1	B_2			
	D_2			E_1	D_3

Weaving Procedure: $A_2 \rightarrow A_1$; $B_2 \rightarrow B_1$; $C_2 \rightarrow C_1$; $E_2 \rightarrow E_1$; $D_3 \rightarrow D_2$; $D_2 (D_3) \rightarrow D_1$; $F_3 \rightarrow F_2$; slide E_3 in the slot so that $E_3 \rightarrow E_1 (E_2)$ and $F_2 (F_3) \rightarrow F_1$.

The Size of Things

If you were asked to identify the largest building in the area where you live, how would you respond? Deciding which of the buildings in your area is largest might be easy. If all the buildings in the United States were included, it may be difficult to decide. The Sears Tower in Chicago might be a candidate since it has a height (without antenna) of 1,454 feet. The Empire State Building in New York City has a height (without lightning rod) of 1,224 feet, but it could be considered since it has 2,248,370 square feet of rentable space.

Sears Tower

Empire State Building

The Boeing aircraft assembly plant in Everett, Washington, could be considered since it contains 472 million cubic feet of space. However, its maximum height is only 115 feet.

Think About This Situation

Length, *area*, and *volume* are three measures often used to describe the size of a space-shape.

a What measure was used to describe the size of the Sears Tower? The Empire State Building? The Boeing assembly plant?

b When describing the size of a space-shape, how do you know what measure to use?

c How is the unit of measure determined by this choice?

INVESTIGATION 1 Describing Size

Length, perimeter, area, and volume are ideas used to describe the size of geometric shapes. These measures commonly are used in daily life as well as in manufacturing and technical trades. This investigation will help you review and deepen your understanding of these basic measurement ideas.

1. Obtain measuring equipment from your teacher and individually select a box such as a cereal or laundry detergent box.

   a. Make a sketch of your box. What technique did you choose? Why?

   b. Working with a partner, compare your boxes. Which do you think is larger?

   c. What characteristics of the boxes did you use to predict which was larger?

2. Working individually, measure your box in as many different ways as you can. Use your measurements to answer each of the following questions.

   a. What is the total length of all the edges? Describe the procedure you used to determine this total. Is there a shortcut? If so, describe it.

   b. Suppose the box could be formed by simply attaching the faces along their edges. How much material would be needed to make it? Describe how you found your answer and note any shortcuts taken.

   c. About how many sugar cubes could your box hold? How did you find your answer?

   d. With your partner, review the prediction you made in Part b of Activity 1. Do all your measures in this activity support your choice? Explain why or why not.

3. Now connect your work in Activity 2 to the sketch of the box you prepared in Activity 1.

   a. On your sketch of the box, write only the lengths that are *absolutely* necessary to answer Parts a through c of Activity 2.

   b. Explain how the lengths you wrote are used to obtain the measures asked for in Activity 2.

   c. If you used any formulas in completing Activity 2, write them down. Then explain how each formula works.

The number you were asked to find in Part b of Activity 2 is the *surface area* of your box. The **surface area** of a space-shape is the sum of the areas of each face—top, bottom, and *lateral* faces. In Activity 2, Part c, you were asked to find the *volume* of your box.

4. The perimeter and area of some plane-shapes are not as easy to find as those for the faces of the boxes. Shown below are several plane-shapes drawn on a background grid. The grid squares have a side length of 0.5 cm.

   a. Find the perimeter of each plane-shape to the nearest 0.5 centimeter. Divide the work among your group. Which shape has the greatest perimeter?

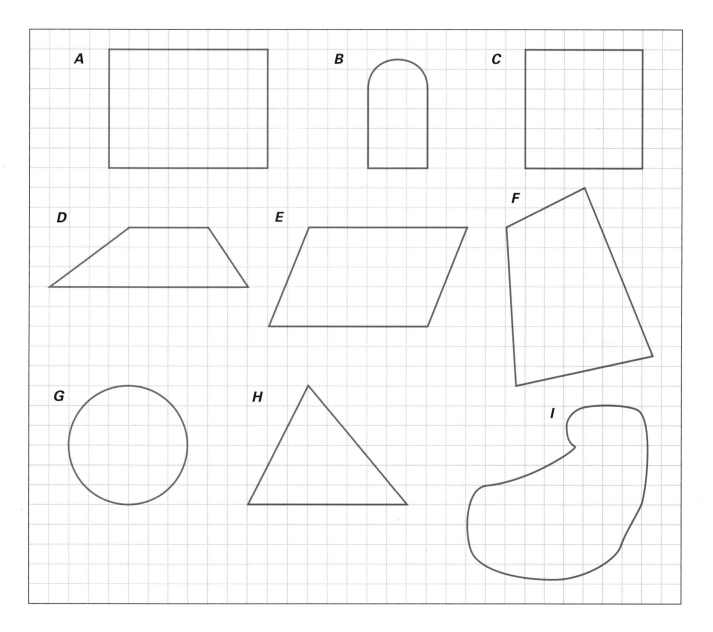

b. Describe the procedure you used to determine each perimeter. If you used any formulas to help you, include them and describe how they were used.

c. Of the procedures you identified in Part b, could any be used for all the plane-shapes? If so, which ones? Why don't the other procedures work for all shapes?

The area of a plane-shape can be measured in different ways. One way is to find the number of same-sized square regions needed to completely cover the enclosed region.

5. Refer back to the plane-shapes in Activity 4.

a. Explain why the area of each grid square is 0.25 square centimeter.

b. For each of the plane-shapes, estimate its area to the nearest 0.25 square centimeter. Again, divide the work among the members of your group. Describe the method you used to determine the area of each region. If you used formulas, note them in your descriptions and explain how they were used.

c. Which of your methods will work for all plane regions? Why won't the other methods work for all regions?

Checkpoint

Often you can find the perimeter and area of a plane-shape in more than one way.

ⓐ Describe two ways in which you can find the perimeter of the parallelogram shown here. Use each method to find the perimeter.

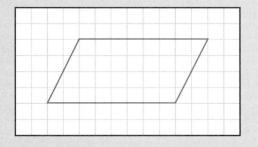

ⓑ Describe two ways in which you can find the area of the parallelogram. Find the area using both methods.

Be prepared to explain to the class the procedures used by your group and the results you obtained.

> ### On Your Own
>
> Consider this scale drawing of Mongoose Lake. Using the given scale, estimate the perimeter to the nearest 10 meters and the area to the nearest 100 square meters.

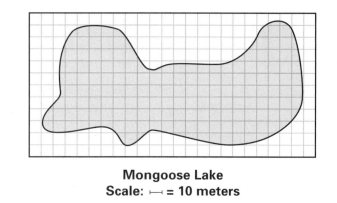

Mongoose Lake
Scale: ⊢—⊣ = 10 meters

Six formulas used to calculate the perimeter and area of common plane shapes are summarized below. You should be familiar with these from your study of mathematics in middle school.

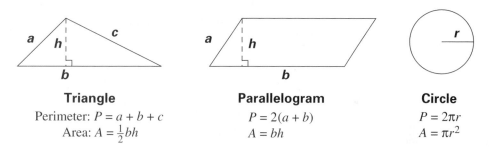

Triangle	**Parallelogram**	**Circle**
Perimeter: $P = a + b + c$	$P = 2(a + b)$	$P = 2\pi r$
Area: $A = \frac{1}{2}bh$	$A = bh$	$A = \pi r^2$

6. Examine the perimeter and area formulas above.

   a. For each formula, what is the meaning of the letters a, b, c, h, or r?

   b. Use the formulas above to compute the areas of figures A, C, E, G, and H of Activity 4. Recall that each grid square has side length 0.5 cm.

   c. Compare your results in Part b with those you obtained for Part b of Activity 5.

   d. Illustrate how you could use the formulas above to compute the areas of figures B and D of Activity 4. Compare your results with those obtained previously.

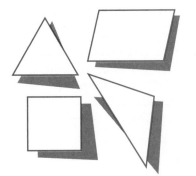

7. Now examine more closely the formulas for the perimeter and area of a triangle and of a parallelogram.

   a. Write a formula for the perimeter of an equilateral triangle that is different from the formula for a general triangle. Write a similar formula for the perimeter of a square.

   b. If you only remember the area formula for a parallelogram, how could you figure out the area formula for a triangle?

   c. How could you modify the formula for the area of a parallelogram so that it applies only to squares? Illustrate with a sketch.

8. In your group, investigate whether each of the following statements is true.

   ■ If two rectangles have the same perimeter, then they must have the same area.

   ■ If two rectangles have the same area, then they must have the same perimeter.

   a. Begin by dividing your group into two approximately equal subgroups, A and B.

   Subgroup A: Develop an argument supporting the first statement or provide a counterexample. (A **counterexample** is an example showing that a statement is not true.) Use rectangles with the same perimeter (such as 40 cm) and make a data table showing length, width, and area.

   Subgroup B: Develop an argument supporting the second statement or give a counterexample. Use rectangles with the same area (such as 48 cm²) and make a data table showing length, width, and perimeter.

   b. Share and check the findings of each subgroup. Then compare your group's final conclusions with those of other groups.

   c. Would your group's views on the two statements be the same or different if "rectangles" were replaced by "squares"? Explain.

9. Jacob has 30 meters of fencing to enclose a garden plot in the shape of a rectangle.

   a. Find the dimensions of all possible gardens Jacob could make using whole-number sides.

   b. Find the areas of the gardens in Part a. Put all your information in a table. Of your sample garden dimensions, which give the largest garden?

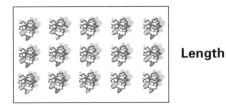

Length

Width

c. Let x represent the width of a garden whose perimeter is 30 meters.

- Write an expression for the length of the garden.

- Write an expression for the area of the garden.

d. Use your graphing calculator or computer software to find the dimensions of the largest rectangular garden that can be enclosed with 30 meters of fencing. Find the dimensions to the nearest 0.1 meter.

e. Did you use tables of values or graphs in Part d? Verify your answer to Part d using the other form of representation.

f. Suppose Jacob has 75 meters of fencing. What is the largest rectangular garden he can enclose? Support your position.

g. What is similar about the shape of the two largest garden plots in this activity?

Checkpoint

Look back at your work calculating perimeters and areas of rectangles.

a If two rectangles have equal areas, can you conclude anything about their perimeters? If so, what?

b If two rectangles have equal perimeters, can you conclude anything about their areas? If so, what?

c For a given perimeter, can you say anything about the shape of the rectangle having the largest possible area? If so, what?

Be prepared to share and defend your group's conclusions.

▶ On Your Own

Consider possible rectangles with a perimeter of 126 units.

a. How many different rectangles with whole number dimensions and a perimeter of 126 units can you create?

b. Of the rectangles in Part a, which has the maximum area?

c. To the nearest 0.1 unit, find the dimensions of the rectangle with a perimeter of 126 units that has the largest possible area.

INVESTIGATION ▶2 Television Screens and Pythagoras

Television manufacturers often describe the size of their rectangular picture screens by giving the length of the diagonal. The set pictured here has a 25-inch diagonal screen. Several companies also advertise a 50-inch diagonal color stereo television. How well does giving the measure of the diagonal describe the rectangular screen?

1. For this activity, consider a 20-inch TV picture screen.

   a. Model the 20-inch diagonal by drawing a 5-inch segment on your paper. (Each member of your group should do this.) In your drawing, each inch represents 4 inches on the picture screen. This is done so that the drawing will fit on your paper. Your drawing has a 1 to 4 (1:4) scale.

   b. Draw a rectangle with the segment you drew as its diagonal. One way to do this is to place a piece of notebook paper with a 90° corner over the segment. Carefully position the paper so that its edges just touch the ends of the segment. Mark the corner. Describe how the entire rectangle can be drawn from this one point and the segment.

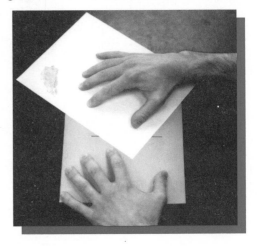

   c. Compare your rectangular screen with those of others in your group. Are they the same? How do they compare in perimeter? In area?

   d. Draw a scale model of a TV screen that is 4 inches wide and has a diagonal of 20 inches. What are the lengths of the other sides? What is its area?

   e. Make a scatterplot of at least eight (*width, area*) data pairs for 20-inch diagonal screens. Use your plot to estimate the 20-inch diagonal screen with the greatest viewing area.

   f. How well does the length of the diagonal of a TV screen describe the screen? Give reasons for your opinion.

2. A manufacturer of small personal TVs thinks that a screen measuring 6 inches by 8 inches is a nice size for good picture quality. What diagonal length should be advertised?

3. You can calculate the diagonal lengths of screens by considering the right triangles formed by the sides. To see how to do this, examine the figures below in which squares have been constructed on the sides of right triangles.

= 1 square unit

a. For each right triangle, calculate the areas of the squares on the triangle's sides. Use the unit of area measure shown. To calculate areas of squares on the longest side, you may have to be creative as suggested by the first figure. Record your data in a table like the one below.

	Area of Square on Short Side 1	Area of Square on Short Side 2	Area of Square on Longest Side
i.	_____	_____	_____
ii.	_____	_____	_____
iii.	_____	_____	_____

b. Describe any pattern you see in the table. Check to see if the pattern holds for other right triangles. Have each member of your group draw a different test case on a sheet of square dot paper. Record these data in your table as well.

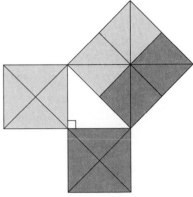

c. Now examine the figure at the right. Explain how this figure illustrates the general pattern in the table of data you prepared in Part b.

d. In general, how are the areas of the squares constructed on the two shorter sides of a right triangle related to the area of the square on the longest side? Compare your discovery with those made by other groups.

e. Represent the lengths of the two shorter sides of the right triangle by *a* and *b*. How could you represent the area of each of the two smaller squares? If the length of the longest side is *c*, what is the area of the largest square?

f. Write an equation expressing the conjecture you made in Part d using *a*, *b*, and *c*.

The discovery you made in Activity 3 was based on a careful study of several examples. The Greek philosopher Pythagoras is credited with first demonstrating that this relationship is true for all right triangles. The relationship is called the **Pythagorean Theorem**. Special cases of this relationship were discovered earlier and used by the Babylonians, Chinese, and Egyptians. A visual proof of the Pythagorean Theorem is suggested in Organizing Task 1 on page 368 and is examined in Extending Task 1 on page 371.

4. Now investigate how the Pythagorean Theorem can help in sizing television screens.

a. Recall the 6-by-8 inch TV screen.

- Find the square of the diagonal.

- How can you find the length of the diagonal when you know its square? What do you get in this case?

- How do you think the manufacturer advertises the size of the 6-by-8 inch TV set? Compare this answer to what you proposed in Activity 2.

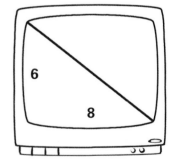

b. Use your calculator or computer to find the length of a diagonal whose square is 56.

c. In Activity 1, the diagonal of the TV screen was 20 inches. Suppose one of the sides of the screen is 13 inches. Find, to the nearest 0.1 inch, the length of the other side.

d. For a 50-inch TV, find five possible length-width pairs that would be reasonable dimensions for a rectangular screen.

5. In Part f of Activity 3, you wrote an equation relating the squares of the lengths of the sides of a right triangle. Jonathan wondered if that *Pythagorean relation* is true for triangles without a right angle. Draw several triangles, measure the sides, and test the Pythagorean relation. Does it work for such triangles? Divide the work among your group and summarize your findings.

6. Now consider cases of other triangles for which the Pythagorean relation does hold.

a. Verify that the numbers 8, 15, and 17 satisfy the Pythagorean relation. Use tools such as a compass or pieces of uncooked spaghetti to draw a triangle whose lengths, in centimeters, are 8, 15, and 17. What kind of triangle is formed?

b. Find four more sets of three numbers that satisfy the Pythagorean relation. Make triangles with these lengths as sides. Divide the work and report the kind of triangles you get.

Checkpoint

The Pythagorean relation is an important property of, and test for, right triangles.

a Write a calculator keystroke sequence to compute the length of the longest side of a right triangle with shorter sides of lengths 7 cm and 10 cm.

b Write a keystroke sequence to compute the length of the third side of a right triangle when the longest side is 25 cm and one other side is 7 cm.

c Given the lengths of the sides of any triangle, how can you tell if the triangle is a right triangle?

Be prepared to share and defend your group's procedures and right-triangle test.

In a right triangle, the longest side (the one opposite the right angle) is called the **hypotenuse** and the shorter sides are called **legs**. The Pythagorean Theorem can be stated in the following form:

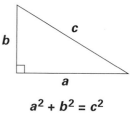

$$a^2 + b^2 = c^2$$

The sum of the squares of the legs of a right triangle equals the square of the hypotenuse.

> ## On Your Own

The Morgan family made a garden for perennial flowers in the corner of their lot. It was a right triangle with legs of 8 and 10 meters.

a. Find the perimeter of the garden and explain how this information might be used by the Morgans.

b. Find the area of the garden and explain how this information might be used in care of the garden.

MORE

Modeling • Organizing • Reflecting • Extending

Modeling

1. A TV manufacturer plans to build a picture screen with a 15-inch diagonal.

   a. Find the dimensions of four possible rectangular picture screens.

   b. Find all picture screens with whole number dimensions. Describe the procedure you used to do this.

   c. Of all possible 15-inch screens, which design would give the largest viewing area? Do you think this would be a good design? Why or why not?

2. The Alvarez family purchased twenty 90-cm sections of fencing to protect their planned garden from the family dog. They plan to use an existing wall as one of the borders. The 90-cm sections can not be cut or bent.

   a. Make a table of the dimensions and areas of gardens that can be enclosed by the 90-cm sections.

   b. What is the largest garden area that can be enclosed with these sections of fence?

   c. For a garden with the largest area, how many sections of fence should be used for each width of the border? For the length of the border?

d. The Alvarez family could have purchased fifteen 120-cm sections of fencing for the same price as the shorter sections they bought. Could they have enclosed a larger garden area using these sections? Explain your response.

e. What else might influence the decision about how to set up the garden?

3. A historical museum plans to paint a mural on its walls illustrating different flags of the world. The flags of the Czech Republic, Switzerland, Thailand, and Japan use some combination of red, white, or blue as shown below. Each of the flags is to be 3 yards long and 2 yards wide. A quart (32 ounces) of paint covers approximately 110 square feet. Paint can be purchased in cans of 32, 16, 8, and 4 ounces. How much paint of each color should be purchased to paint these flags?

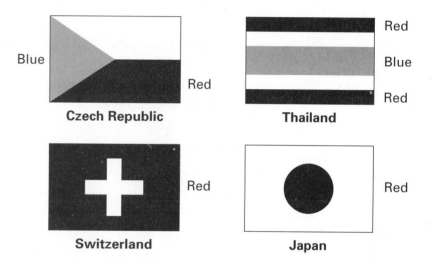

4. Materials tend to expand when heated. This expansion needs to be considered carefully when building roads and railroad tracks. In the case of a railroad track, each 220-foot-long rail is anchored solidly at both ends. Suppose that on a very hot day a rail expands 1.2 inches, causing it to buckle as shown below.

a. At what point along the rail do you think the buckling will occur?

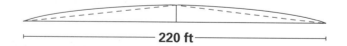

b. Do you think you could slide a gym bag between the raised rail and the track bed?

c. Model this situation using right triangles, and then calculate an estimate of the height of the buckle.

d. Would you expect your estimate of the height of the buckle to be more or less than the actual value? Explain your reasoning.

e. Research *expansion joints*. How does the use of these joints in railroad tracks and concrete highways minimize the problem you modeled in Part c?

5. In Lesson 1, Investigation 1, you used an 11-inch-long sheet of paper to model a rectangular prism column.

a. If 0.5 inch is used for overlapping before taping the seam, how should the paper be folded to make the area of the rectangular base as large as possible?

b. Does the paper folded as in Part a produce the rectangular prism column with the largest possible surface area? Organize your work and provide evidence to support your answer.

Organizing

1. Carefully trace the figure shown here.

a. Cut out the square labeled *A* and the pieces labeled *B*, *C*, *D*, and *E*.

b. Can you use the five labeled pieces to cover the square on the hypotenuse of the right triangle? If so, draw a sketch of your covering.

c. This puzzle was created by Henry Perigal, a London stockbroker who found recreation in the patterns of geometry. How is Perigal's puzzle related to the Pythagorean Theorem?

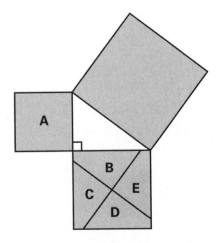

d. After studying the puzzle, Anne conjectured: "Every square can be dissected into five pieces which can be reassembled to form two squares." Do you think Anne is correct? Explain your reasoning.

2. The television industry has set a standard for the sizing of regular television screens. The ratio of height h to width w, called the *aspect ratio*, is 3:4. That is $\frac{h}{w} = \frac{3}{4}$.

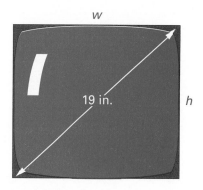

   a. Write an equation expressing h as a function of w.

   b. Use the Pythagorean Theorem to write an equation relating h, w, and the diagonal length 19.

   c. Use your equations in Parts a and b to find the standard dimensions of a 19-inch TV screen.

   d. Check the dimensions you obtained against actual measurements of a 19-inch TV screen.

3. Draw squares of side lengths 2, 4, 7, 8, 10, and 11 centimeters on centimeter grid paper.

   a. Measure the diagonals to the nearest 0.1 cm. Record your data in a table.

   b. Make a plot of your (*side length*, *diagonal*) data.

   c. Does the plot appear to have a linear pattern? If so, find a linear model that you believe fits the trend in those data.

   d. Find an equation for your line.

   - What is the slope? What does it mean?

   - What is the *y*-intercept? Does it make sense? Explain.

   e. Predict the length of the diagonal of a square with side length of 55 cm.

   f. Compare your predicted length to that computed by using the Pythagorean Theorem. Explain any differences.

4. Locate six cylindrical shapes of different sizes. Cans and jar lids work well.

   a. Measure the circumference of each cylinder to the nearest 0.1 cm. Record your data in a table.

   b. On a sheet of paper, trace around the base of each cylinder. Then measure, to the nearest 0.1 cm, the diameter of each tracing and record it in your table.

   c. Make a plot of your (*diameter*, *circumference*) data.

    d. Find a line and its equation that you believe models the trend in these data.

        ■ What is the slope of the line? What does it mean?

        ■ What is the y-intercept? Does it make sense? Explain.

    e. The diameter of a small fruit-juice can is approximately 5.5 cm. Use your linear model to predict the circumference of the can.

    f. Compare your predicted circumference to that computed by using the formula for the circumference of a circle. Explain any differences.

5. Imagine a plane intersecting a cube.

    a. Describe the shape of the intersection of a cube and a plane that has the largest possible area.

    b. Describe the shape of the intersection of a cube and a plane that has the smallest possible area.

    c. Find the least and greatest possible areas of shapes formed by the intersection of a plane with a cube 5 cm on a side.

6. Draw a segment 10 cm long on a sheet of paper.

    a. Imagine a right triangle that has the segment as its hypotenuse. If one side is 0.5 cm long, where would the vertex of the right angle be? Locate that point and make a mark there. (Use a technique similar to the one in Activity 1 of Investigation 2, page 362.) Where would that vertex be if the side is 1 cm long? Mark that point. Now increase the side length in steps of 0.5 cm and plot those vertices of the resulting triangles. Stop when the side length is 10 cm.

    b. Examine your plot of the points. What shape do they appear to form?

    c. Support your view by citing appropriate measurements from your model.

Reflecting

1. Architects use many design principles. For example, tall buildings will always provide more daylight, natural ventilation, and openness than low buildings of the same floor area. Explain why this is the case.

New York City

2. Which shape encloses more area: a square with perimeter 42 or a circle with perimeter (circumference) 42? Justify your reasoning.

3. In what kind of units is area measured? How can you use this fact to avoid confusing the formulas $2\pi r$ and πr^2 when computing the area of a circle?

4. Experimenting, collecting data, and searching for patterns is a powerful way to discover important mathematical relationships. However, it also has limitations. Consider the following example.

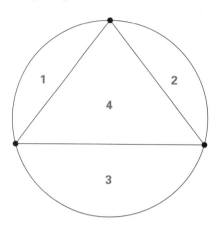

On the circle at the right, three points are marked. Each point is connected to all the others. Four nonoverlapping regions are formed.

Anthony investigated what happened when differing numbers of points were marked on a circle and each was connected to all the others. He summarized his findings in the table below.

Number of Points	1	2	3	4	5
Number of Regions	1	2	4	8	16

a. Do you see any pattern in the table?

b. Anthony predicted that the number of nonoverlapping regions formed by six points would be 32. Check his prediction.

c. What lesson can be learned from this example?

Extending

1. Examine more closely the *dissection* method of "proof" of the Pythagorean Theorem in Organizing Task 1.

a. In general, to determine one of the "cut lines," position a copy of the right triangle so that its hypotenuse goes through the center of the $a \times a$ square as shown. How is the second "cut line" related to the first? Draw both "cut lines" on a copy of the figure shown. Label the remaining lengths of the sides of the four pieces in the $a \times a$ square.

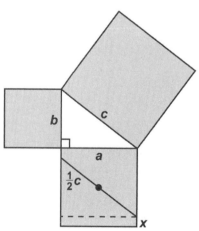

b. Using your labeling of the segments, show that when the four pieces of the $a \times a$ square are rearranged in the $c \times c$ square, $c^2 = a^2 + b^2$.

2. The circle at the right has been dissected into eight sections. These sections can be reassembled to form an "approximate" parallelogram.

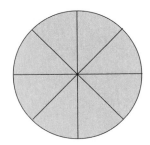

a. How is the base of this "approximate" parallelogram related to the circle?

b. What is the height of the "approximate" parallelogram?

c. How could you dissect the circle into sections to get a better approximation of a parallelogram?

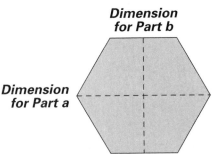

d. Use the above information to produce the formula for the area of a circle.

3. Suppose you are designing a hexagonal flower garden for which all sides are the same length and all angles are the same size.

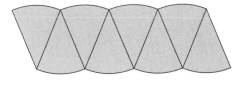

a. How much decorative fencing is needed to enclose the garden if it measures 4 meters between opposite vertices?

b. Suppose the distance between opposite *sides* is to be 4 meters. How much fencing is needed?

c. How many flowers can be planted in the garden described in Part a if each flower requires 225 square centimeters?

d. How many flowers can be planted in the garden described in Part b?

e. Suppose the 4-meter length measured the length of other segments through the center of the hexagon. What can you say about the amount of fencing needed? Explain your reasoning.

4. In order for kites to fly well, they need to have a high ratio of *lift area* to weight. For two-dimensional kites, the lift area is just the area of the kite. Find the lift area of the traditional kite shown with cross pieces of lengths 0.8 m and 1.0 m.

INVESTIGATION ▶ 3 Size Measures for Space-Shapes

Prisms model the shape of most office buildings. Among the prisms, the rectangular prism is the most common. This fact is demonstrated in an aerial photo of a portion of New York City, shown below at the left. Other prisms well suited to architecture generally have numbers of sides that are multiples of four. One building in the Crown Zellerbach Plaza in San Francisco (shown below at the right) is based on a 40-gon prism. (A 40-gon is a polygon with 40 sides.) Regardless of the shape of the base, the other faces (*lateral faces*) of a prism are rectangles. These facts permit easy measuring of the surface area and volume of space-shapes commonly found in architecture.

1. In your group, discuss possible real-life situations where you would want to know:

   a. the areas of the lateral faces of a prism;

   b. the surface area (total area of the bases and lateral faces) of a prism;

   c. the volume of a prism.

2. Now try to discover a formula for calculating the total area of the lateral faces of a prism.

   a. Refer to the skeleton models of the five prisms that you made in Activity 8 of Investigation 2 (page 333). Imagine covering each prism with paper. Describe how you would find the total area of the lateral faces of each of those prisms.

   b. Based on your descriptions in Part a, write a general formula for finding the total area of the lateral faces of any prism. Describe what each variable in your formula represents.

c. Suppose you have an octagonal prism whose bases have sides of length 4 cm and all angles the same size. The edges connecting the bases are 10 cm long. Apply your formula to this prism.

d. Simplify your formula in Part b for prisms with an equilateral, equiangular base. Could you use the concept of perimeter to further simplify your formula? If you can, do so; if not, explain why not.

e. How could you simplify your formula for the special case of a cube?

3. Modify your formulas of Activity 2 Parts b and d to give formulas for the *surface area* of the prisms.

a. Find the surface area of one of the boxes you analyzed in Activity 1 of Investigation 1 (page 356).

b. Find the surface area of each of the following prisms.
- Square prism: 5-cm edges on bases, height is 6 cm.
- Triangular prism: 5-cm edges on bases, height is 6 cm.

c. The base of a hexagonal prism with 4-cm edges and all angles the same size can be divided into 6 equilateral triangles as shown here.

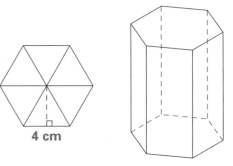

4 cm

- How can you calculate the altitude of one of the equilateral triangles?
- Calculate the area of a base.
- Find the surface area of this prism if the height is 9 cm.

Surface area gives you one measure of the size of a space-shape. *Volume* gives you another. Just as area can be found by counting squares (hence *square units*), volume can be found by counting cubes (hence *cubic units*).

4. Look at the figure below showing one layer of unit cubes in a prism.

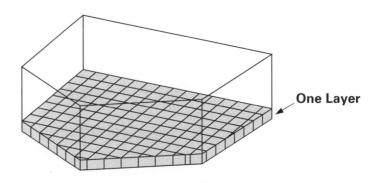

One Layer

a. If that layer has a base area of 84 square units, how many unit cubes are in it?

b. If there are 6 such layers, what is the volume of the prism? What is the volume if there are 134 layers?

c. Suppose the layer has a base area of 13 square units. How many unit cubes are in it? If there are *h* such layers, what is the total volume?

d. Write a formula for calculating the volume *V* of a prism which has a base area of *B* square units and a height of *h* units. Compare your formula with that of other groups. Resolve any differences.

5. Make a sketch of each prism with the given dimensions and then find its volume.

   a. Rectangular prism: base is 4 cm by 5 cm, height is 6 cm.

   b. Cube: edge is 8 cm.

   c. Equilateral triangular prism: base edges are 3 cm, height is 8 cm.

   d. Equilateral, equiangular hexagonal prism: base edges are 6 cm, height is 3 cm.

6. Find the surface area of each prism in Activity 5.

7. Some products like vegetables or fruit juice are packaged in cylindrical cans. A cylinder (shown here) can be thought of as a special prism-like shape.

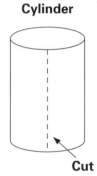

Cylinder

Cut

   a. What shape are its bases? Imagine removing the bases and cutting the lateral face from base to base as indicated in the diagram. If you flatten the cut lateral face, what shape is it?

   b. Find the surface area and volume of a cylinder if the base has a radius of 4 cm and the height is 7 cm.

   c. Write a formula for the surface area *S* of a cylinder with radius *r* and height *h*.

   d. Write an equivalent expression for your formula for surface area. Which expression will be easier to remember?

   e. Write a formula for the volume *V* of a cylinder with radius *r* and height *h*.

8. A can of diced tomatoes has a base with diameter 7.5 cm and height 11 cm. These cans are usually packaged for shipping in boxes of 24 (two layers of 12).

   a. What space-shape usually is used for the shipping boxes?

   b. What arrangements of the cans are possible for a box holding 12 in a layer?

c. For each arrangement of 12 cans, determine the dimensions of the smallest possible shipping box.

d. What arrangement of cans uses the greatest amount of available space in the box?

e. Which arrangement of cans requires the smallest amount of cardboard to make the box?

9. A $10\frac{3}{4}$ ounce can of soup has a height of 10 cm and a base with radius 3.3 cm. A $6\frac{1}{8}$ ounce can of tuna has a height of 4 cm and a base with radius 4 cm. Which metal can is more efficient in its use of metal for the weight of its contents?

Checkpoint

In this investigation, you explored methods and discovered formulas for finding the surface area and volume of prisms and cylinders.

ⓐ How is the perimeter of a base of a prism useful in finding the area of the lateral faces?

ⓑ How are the formulas you developed for surface area and volume of cylinders similar to the corresponding formulas for prisms? How are they different?

ⓒ How is the Pythagorean Theorem helpful in computing surface areas and volumes of prisms with equilateral triangles or equilateral, equiangular hexagons as bases?

Be prepared to share your group's ideas with the entire class.

▶ On Your Own

A decorative candy tin used by Sorby's Candies and Nuts has an equilateral, equiangular hexagonal base with dimensions as shown. These tins are shipped from the manufacturer in boxes 20 cm by 17.5 cm.

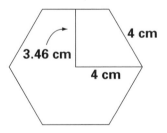

a. How many tins can be put in one layer of a box?

b. If the box and tin have the same height, how much of the available volume of the box is used by the candy tins?

Modeling • Organizing • Reflecting • Extending

Modeling

1. Susan and John Sitzman are farmers who specialize in raising cattle. They need to build a new feed storage silo with an estimated volume of 20,000 cubic feet for the cylinder portion. They have prepared a square piece of land, 20 feet on each side, on which to build the silo.

   a. After shopping around, the Sitzmans found the "best buy" comes from a company which makes a type of silo with a diameter of 18 feet and a height of 55 feet. Would this type fit their needs? Why or why not?

   b. ACME Equipment Company makes silos with a range of diameters from 16 to 21 feet. They will custom-build their silos to any height the customer desires. If the Sitzmans require a volume of 20,000 cubic feet, how high would a silo of a given diameter have to be? Make a table like the one below to help organize your work.

Diameter (in feet)	16	17	18	19	20	21
Height (in feet)						

   c. The Sitzmans want to keep the height of their new silo below 75 feet, since they often have strong winds in their area. Which of the silos in your table would be suitable for these conditions?

2. A computer disk box measures 9.7 cm by 4.8 cm by 9.7 cm. Inside, the bottom half of the box is an open-ended cardboard protective liner which measures 7.2 cm high at its back and 5.8 cm high at its front. The thickness of all material is 0.1 cm.

   a. Sketch the liner and give its dimensions.

   b. How much cardboard is needed to make the disk box and its liner?

   c. What is the volume of the box?

   d. What is the volume of the liner?

3. A container manufacturing company makes open-top storage bins for small machine parts. One series of containers is made from a square sheet of tin that is 24 cm on a side. Squares of equal size are cut from each corner. The tabs are then turned up and the seams are soldered.

 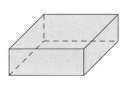

a. Using centimeter graph paper, cut out two squares 24 cm on a side. Make model bins by cutting squares of one and two centimeters on a side from the corners of the 24 × 24 squares. Fold up the tabs and tape the seams. What are the dimensions of the model which has the larger volume?

b. Let x represent the side-length of the cutout corner squares. Write an expression
  ■ for the length of each side of the storage bin;
  ■ for the volume of the bin.

c. What is the possible range of values for x, the side-length of the cutout squares?

d. Use the table-building capability of your graphing calculator or computer software to help find the dimensions of the container with the largest possible volume.

4. A swimming pool is 28 feet long and 18 feet wide. The shallow 3-foot-deep end extends for 6 feet. Then for 16 feet horizontally, there is a constant decline toward the 9-foot-deep end.

a. Sketch the pool and indicate the measures on the sketch. Is this a prism? If so, name it.

b. How much water is needed to fill the pool within 6 inches of the top?

c. One gallon of paint covers approximately 75 square feet of surface. How many gallons of paint are needed to paint the inside of the pool? If the pool paint comes in 5-gallon cans, how many cans should be purchased?

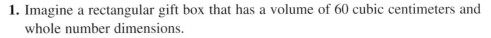

d. How much material is needed to make a pool cover that extends 2 feet beyond the pool on all sides?

e. About how many 6-inch square ceramic tiles are needed to tile the top 18 inches of the inside faces of the pool?

Organizing

1. Imagine a rectangular gift box that has a volume of 60 cubic centimeters and whole number dimensions.

  a. Find the dimensions of all possible boxes. Use a table to organize your findings.

  b. What is the surface area of each gift box?

  c. What are the dimensions of the gift box that uses the least amount of paper?

  d. Express your view on the statement: *If two prisms with the same kind of bases (the same shape, not the same size) have identical volumes, then their surface areas are identical.* Explain your position.

2. To use the formula for the area of a triangle, you need to know the length of a side (the base) and the length of the altitude to that side (the height). In the case of some special triangles, it's sufficient to know the length of a side.

  a. Develop a formula for calculating the area A of an equilateral triangle with side-length s.

  b. Develop a formula for calculating the area A of an equilateral, equiangular hexagon with side-length s.

3. Ian's class had two prisms that were 20 cm high. One had a square base with 5-cm sides. The other had an equilateral triangular base with 5-cm sides. The class collected data on the amount of water needed to raise the water level to various heights in each prism. The data are summarized in the following table.

Height in cm	0	2	4	6	8	10	12	14	16	18	20
Volume of Square Prism (5-cm sides)	0	49	102	150	199	252	300	347	398	452	500
Volume of Triangular Prism (5-cm sides)	0	21	44	65	86	107	130	152	172	196	216

  a. Produce a scatterplot of the (*height, volume*) data for each prism.

  b. For which scatterplots would a linear model fit the trend in the data? Where appropriate, find an equation for the linear model.

  c. Describe the rate of change in the volume for each prism.

  d. How is the rate of change related to the base of the space-shape?

4. Ian's class found the results of modeling the volumes of the prisms being filled with water (Organizing Task 3) very interesting. They extended the investigation and collected (*height, volume*) data for other 20-cm-tall space-shapes. The results are summarized in the following table.

Height in cm	0	2	4	6	8	10	12	14	16	18	20
Volume of Square Pyramid (5-cm sides)	0	17	33	50	66	84	99	115	133	149	167
Volume of Triangular Pyramid (5-cm sides)	0	7	16	22	28	36	44	50	57	66	72
Volume of Cylinder (2-cm radius)	0	25	49	75	101	123	149	177	200	227	252
Volume of Cone (2-cm radius)	0	8	17	26	34	41	49	59	66	76	84

a. Produce four scatterplots of the (*height, volume*) data in the table.

b. If appropriate, find a linear model for each scatterplot.

c. Compare the linear models for the cone and the cylinder. Make a conjecture about the relationship between the volumes of a cylinder and a cone with identical bases and identical heights.

d. Compare the linear models for the pyramids with those of a prism with a similar base in Organizing Task 3. Make a conjecture regarding the relation between the volumes of a prism and a pyramid with identical bases and identical heights.

Reflecting

1. Commodities sold in grocery stores come in many kinds of packages. Packages made of cardboard tend to be rectangular prisms. Packages made of metal or glass are more often cylinders. What reasons might you give for these trends?

2. What changes the volume of a cylinder more: doubling its diameter or doubling its height? Explain your reasoning.

3. In Investigation 3, you saw that the volume of a prism or a cylinder is found simply by multiplying the area of the base by the height. Explain why it would not be reasonable to calculate volumes of pyramids or cones in the same way.

4. In Unit 4, "Graph Models," you developed *algorithms* for finding Euler circuits and coloring the vertices of a graph. In this investigation, you developed *formulas* for calculating volumes of space-shapes. How is a formula similar to an algorithm? How is it different?

Extending

1. The U.S. Postal Service has requirements on the size of packages it will ship within the United States. The maximum size for parcel post packages is 108 inches in combined length and *girth*.

   a. Suppose you are shipping goods via parcel post that do not require a specific length shipping box. What are the dimensions of a square prism package that would allow you to ship the greatest volume of goods?

   b. Could you ship a greater volume of goods using a rectangular prism package? Explain your reasoning.

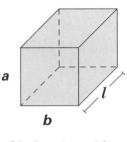

Girth = 2a + 2b

2. A cone has a circular base with a radius of 4 cm. To cut it open, you need to make a 12 cm cut from bottom to top.

   a. Sketch the shape of the cone when cut and flattened.

   b. The flattened shape is part of what larger shape? What fractional part of the larger shape is this surface?

   c. Find the area of the surface of the cone.

3. The diagram at the right suggests that a cone with the same base and height as a cylinder will have less volume. To see how much, use a cylindrical can or glass and a cone made from stiff paper as shown. Fill the cone with sand, rice, or birdseed and then pour the contents into the cylinder. Repeat until the cylinder is filled.

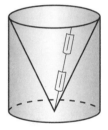

   a. What is the relationship between the volume of the cone and the volume of the cylinder?

   b. Compare your finding with that of other classmates who completed this experiment

   c. Write a formula for the volume V of a cone with radius r and height h.

4. The volume formulas in this unit can be viewed as special cases of the *prismoidal formula*:

$$V = \frac{B + 4M + T}{6} \cdot h$$

where B is the area of the cross section at the base, M is the area of the cross section at the "middle," T is the area of the cross section at the top, and h is the height of the space-shape.

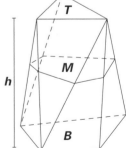

   a. Explain why $\frac{B + 4M + T}{6}$ can be thought of as the "weighted" average area of the cross sections of the solid. Compare this formula for "weighted average" with the one used in Unit 4, "Graph Models," page 313.

b. Use the prismoidal formula and the diagram below to show that the volume of a triangular prism is $V = Bh$, where B is the area of the base and h is the height.

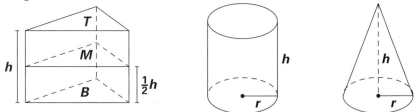

c. Imagine a prism with an equilateral, equiangular base. As the number of sides of the base increases, the shape of the prism approaches that of a cylinder. Use this fact, the cylinder shown above, and the prismoidal formula to show that the volume of a circular cylinder with radius r and height h is given by $V = \pi r^2 h$.

d. Imagine a pyramid with an equilateral, equiangular base. As the number of sides of the base increases, the shape of the pyramid approaches that of a cone. Use this fact, the cone shown above, and the prismoidal formula to help you discover a formula for the volume of a cone with base radius r and height h.

e. Locate an irregularly shaped vase or bottle. Use the prismoidal formula to approximate the volume of the container. Check your estimate by filling the container with water, then pouring it into a measuring cup.

5. Analysis of the formulas for the volume of a prism and for the volume of a cylinder suggests that multiplying the dimensions of the space-shape by a positive constant changes the volume in a predictable way.

a. One large juice can has dimensions twice those of a smaller can. How do the volumes of the two cans compare?

b. One cereal box has dimensions 3 times those of another. How do the volumes of the two boxes compare?

c. If the dimensions of one prism are 5 times those of another, how do the volumes compare?

d. If the dimensions of one prism are k times those of another, how do the volumes compare?

6. The dimensions of a model of a building are one-hundredth of the dimensions of the actual building.

a. How does the volume of the model compare to the volume of the actual building?

b. How does the surface area of the model compare to the surface area of the building?

The Shapes of Plane Figures

Space-shapes come in a wide variety of forms. Many of those forms are based on some very common shapes such as the rectangular prism, the pyramid, the cylinder, and the cone. All the faces of prisms and pyramids, as well as the bases of cylinders and cones, are *plane-shapes*. The most common shape of a face is the *polygon*. In this lesson, you will study the characteristics of polygons and explore how they are used in art, design, and other ways.

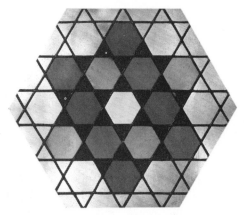

© 1977 Stained Glass Tessellation. Creative Publications Inc. Reprinted by permission of Creative Publications.

Think About This Situation

Examine the portion of a stained glass window shown above.

a What is most visually striking about the pattern?

b What shapes make up the pattern?

c What symmetry is evident in the pattern?

d Some seams appear to be parallel. Why is this?

e Does there appear to be any gaps or overlaps in the tile-like pattern? What might explain your observation?

INVESTIGATION 1 Polygons and Their Properties

The word "polygon" comes from Greek words *poly* meaning "many" and *gon* meaning "angle." This is descriptive, but it describes shapes that are not polygons as well as those that are. It will be helpful to have a more precise definition of "polygon" before investigating properties of various polygons.

1. Shown below are twelve plane-shapes.

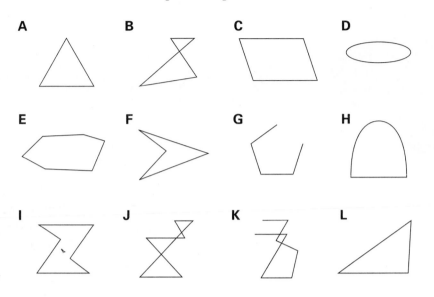

A B C D

E F G H

I J K L

a. Working individually, sort these shapes into groups of shapes that are "alike" in some manner.

b. Share with your group the criteria you used to sort the twelve shapes.

c. Choose another set of sorting criteria for sorting the shapes. Re-sort and describe the criteria you used.

d. As a group, determine a criterion for sorting which would put all shapes except shapes D and H in one group.

e. What criterion would put shapes A, B, C, E, F, I, J, and L (no others) into one group?

f. What criterion would put shapes A, C, E, F, I, and L (no others) into one group?

g. What criterion would put shapes A, C, E, and L (no others) into one group?

h. What criterion would put shapes A and L in one group; B, C, F, and G in another group; and E, I, and J in a third group?

i. What criterion would put shapes A, D, G, and H (no others) into one group?

j. What criterion would put shapes A, C, and D (no others) into one group?

Your work in Activity 1 suggests that there are many ways that plane-shapes can be grouped or classified. One commonly used classification puts *polygons* into one group and all other shapes into another group (non-polygons). The criterion you developed for Part f above should have identified only shapes which are polygons.

2. Review the shapes that were grouped together in Part f of Activity 1.

    a. Are there shapes in that group that did not fit your idea of a polygon? If so, which ones?

    b. What is it about these shapes that made you think they were not polygons?

    c. Draw a shape, either a polygon or not a polygon. Trade shapes with a partner and have each person identify the shape as a polygon or not. Explain why that choice was made.

One class of polygons commonly seen in architecture is the class of **quadrilaterals**, which are four-sided polygons.

3. You are familiar with many quadrilaterals from your previous work in mathematics. The diagram at the right shows a general quadrilateral. It has no special properties; just four sides and four angles.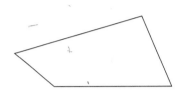

    a. You may recall that there are six special kinds of quadrilaterals. They are the *square*, *rectangle*, *parallelogram*, *rhombus*, *trapezoid*, and *kite*. Draw a sketch of each of these polygons.

    b. For each quadrilateral sketched in Part a, list all the special properties you think it has. For example, are opposite sides parallel or equal in length? Are adjacent angles equal or supplementary (sum of their degree measures is 180)?

    c. Which of your shapes has the greatest number of special properties? Which has the least?

    d. Compare your analysis of properties of special quadrilaterals to those of other groups. Resolve any differences.

Checkpoint

Look back at your work classifying plane-shapes as polygons and as special quadrilaterals.

  a Write a criterion (a definition) that will put all polygon shapes in one group and any other plane-shape in another group.

  b Write criteria you would use to classify a quadrilateral as the following:

    a kite   a trapezoid   a rhombus

    a parallelogram   a rectangle   a square

  Be prepared to compare your group's classification criteria with those of other groups.

Classify each statement as *true* or *false*. Give a justification for your conclusion.

a. Shape A is a polygon.   **b.** Shape B is a polygon.

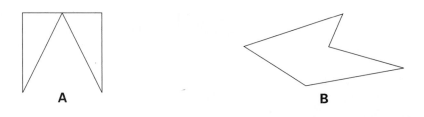

        A   **B**

c. Every parallelogram is a rectangle.

d. Every rhombus is a kite.

e. Every trapezoid is a parallelogram.

A polygon with *n* sides is often called an *n-gon*. So another name for a quadrilateral is a *4-gon*. In the next activity, you will explore connections between the number of sides of a polygon and the measures of its angles.

4. Carefully draw polygons having 4, 5, 6, 7, 8, 9, and 10 sides. Share the work in your group.

    a. Subdivide each polygon into *nonoverlapping* triangles by drawing line segments from one vertex to another. Try to make the *fewest* possible triangles in each polygon.

       ■ Record your results in a table like the one below. Compare the number of sides of a polygon and the number of triangles into which it can be subdivided.

Number of Sides	Number of Triangles
4	

       ■ Examine your table to find a pattern. Draw and test more polygons, if needed.

       ■ Suppose a polygon has *n* sides. Into how many nonoverlapping triangles can it be subdivided? Use the pattern you discovered in your table to write a rule (formula) relating the *number of sides n* to the *number of triangles T.*

b. Recall that the sum of the measures of the angles of a triangle is 180 degrees.

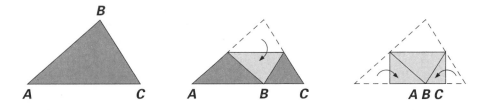

Use this fact to find the sum of the measures of the angles of each polygon listed in your table. Add a column "Angle Sum" to your table and record your results.

- Write a rule relating the number of sides of an *n*-gon to the sum of the measures of its angles.
- Use your rule to predict the sum of the measures of the angles of a 12-gon. Check your prediction with a sketch.

c. Do you think a line is a good model for the (*number of sides, angle sum*) data? If so, what is the slope and what does it mean in terms of the variables? Does the *y*-intercept make sense? Why or why not?

5. Now investigate how to predict the measure of one angle of a **regular polygon**. In a regular polygon, all angles are the same size and all sides are the same length.

   a. Determine the measure of one angle of a regular hexagon. Apply your procedure to a regular 10-gon.

   b. Write a rule relating the *number of sides n* to the *measure A of an angle* of a regular *n*-gon.

6. Polygons and other plane-shapes also may have other properties. For example, look at your grouping rules for Parts i and j of Activity 1 (page 384).

   a. Shapes A, D, G, and H could be grouped together because they each have **reflection** or **line symmetry**. On a copy of each shape, draw a line through it so that one half is the reflection of the other half.

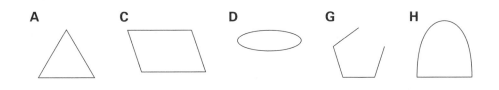

   b. Shapes A, C, and D could be grouped together because they have **rotational symmetry**. This is also called **turn symmetry** because the amount of rotation is given as an angle in degrees. Locate the center of each shape and determine through what angles it can be turned so that it coincides with itself.

c. Which of the following figures have reflection (line) symmetry?

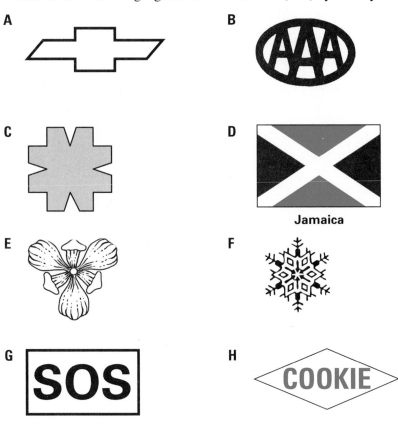

A

B

C

D

Jamaica

E

F

G SOS

H COOKIE

d. Which of the figures in Part c have rotational (turn) symmetry? Estimate the angles through which each can be turned to coincide with itself.

e. In Parts c and d, how did you determine which figures had reflection or rotational symmetry?

f. Which of the quadrilaterals that you drew in Activity 3 (page 385) have reflection (line) symmetry? Which have rotational (turn) symmetry?

g. If a figure has reflection symmetry, must it have rotational symmetry? Explain or give a counterexample.

h. If a figure has rotational symmetry, must it have reflection symmetry? Explain or give a counterexample.

Checkpoint

Regular polygons have predictable angle measures and symmetry properties.

a Does the logo shown at the right have reflection symmetry? Rotational symmetry? If so, trace the logo and sketch lines of symmetry and/or give angles of rotation.

b Describe what is meant by reflection or line symmetry.

c Describe what is meant by rotational or turn symmetry.

d The corner angles of this logo are the same size. Describe how you would determine the measure of any one of these angles without measuring.

Be prepared to share your group's responses with the entire class.

On Your Own

The plane-shape depicted at the right is a drawing of a stop sign (without the word "stop").

a. Is it a polygon? Explain.

b. What is the *sum* of the measures of its angles?

c. Using what you believe to be true about the shape of stop signs, determine the measure of *each* of its angles.

d. What symmetry, if any, does the shape have?

e. Identify another traffic sign that is in the shape of a regular polygon. Describe the shape and its symmetry.

INVESTIGATION 2 Patterns with Polygons

In a regular polygon, all sides are the same length and all angles are the same size. Because of these characteristics, regular polygons have many useful applications. Prism-like packages are often designed with regular polygons as bases. If you think about tiled walls or floors you have seen, it is likely that the tiles were regular polygons.

1. The figures below show portions of **tilings** of equilateral triangles and squares. The tilings are made of repeated copies of a shape placed edge-to-edge. In this way, the tilings completely cover a region without overlaps or gaps.

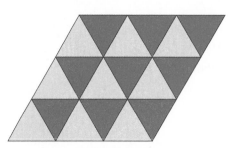 

    a. Assume that the tilings are extended indefinitely in all directions to cover the plane. Describe the symmetries of each tiling.

    b. For each tiling:

- What is the total measure of the angles at a common vertex?
- What is the measure of each angle at a common vertex?

    c. Using the tiling with equilateral triangles, identify other common polygons formed by two or more triangles. Sketch each and show the equilateral triangles that form the shape.

2. You have seen that equilateral triangular regions tile a plane. In this activity, you will explore other kinds of triangular regions that can be used as tiles.

    a. Each member of your group should cut from poster board a small triangle that is *not* equilateral. Each member's triangle should have a shape different from the other members' triangles. Individually, explore whether a tiling of a plane can be made by using repeated tracings of your triangular region. Draw and compare sketches of the tilings you made.

    b. Can more than one tiling pattern be made by using copies of one triangular shape? If so, illustrate with sketches.

    c. Do you think any triangular shape could be used to tile a plane? Explain your reasoning.

3. You saw above that square regions can tile a plane. In this activity, you will explore other quadrilaterals that can be used to make a tiling.

    a. Each member of your group should cut a non-square quadrilateral from poster board. Again, each of the quadrilaterals should be shaped differently. Individually, investigate whether a tiling of a plane can be made with the different quadrilaterals. Draw sketches of the tilings you made.

b. For those quadrilaterals that tile, can more than one tiling pattern be made using the same shape? If so, illustrate and explain.

c. Make a conjecture about which quadrilaterals can be used to tile a plane.

4. You have seen two regular polygons which tile the plane. Now explore other regular polygons that could be used to make a tiling.

a. Can a regular pentagon tile the plane? Explain your reasoning.

b. Can a regular hexagon tile the plane? Explain.

c. Will any regular polygon of more than six sides tile the plane? Provide an argument supporting your view.

Checkpoint

In the first part of this investigation, you explored special polygons that tile the plane.

a Write a summarizing statement describing which triangles and which quadrilaterals tile the plane.

b Which regular polygons tile the plane? Explain your reasoning.

Be prepared to share your group's conclusions and thinking with the entire class.

Tilings that consist of repeated copies of a single regular polygon whose edges exactly match are called **regular tessellations**.

▶ On Your Own

Using a copy of the figure at the right, find a pentagon in the figure that will tile the plane. Shade it. Show as many different tiling patterns for your pentagon as you can.

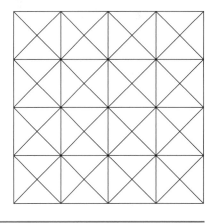

---

5. Tilings can involve more than one type of shape. Two examples of such tilings are shown at the top of the following page. The pattern on the left is from a Persian porcelain painting. The pattern on the right is from the Taj Mahal mausoleum in India.

a. Examine carefully each of these patterns. How many different shapes are used in the tiling on the left? How many different shapes are used in the tiling on the right?

b. Examine the tessellation of regular octagons and squares shown below.

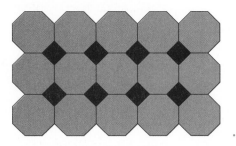

- Use angle measure to explain why the polygons will fit together with no overlaps or gaps.
- At each vertex, is there the same combination and arrangement of polygons?

c. Tessellations of two or more regular polygons that involve the same arrangement of polygons at each vertex are called **semi-regular tessellations**. Test whether a regular hexagon, two squares, and an equilateral triangle can be used to make a semi-regular tessellation. If possible, draw a sketch of the tessellation.

d. Semi-regular tessellations are coded by listing the number of sides of the polygons at each vertex. The numbers are arranged in order with the smallest number first. The tessellation in Part b is 4, 8, 8. Use this code to describe:

- the tessellation of equilateral triangles and regular hexagons at the beginning of this lesson (page 383);
- the tessellation you drew in Part c;
- each of the three possible regular tessellations.

A **net** is a two-dimensional pattern which can be folded to form a three-dimensional shape. R. Buckminster Fuller, inventor of the geodesic dome (see Extending Task 4 on page 353), created a net for a "globe" of the earth. In the modified version below, each equilateral triangle contains an equal amount of the Earth's surface area.

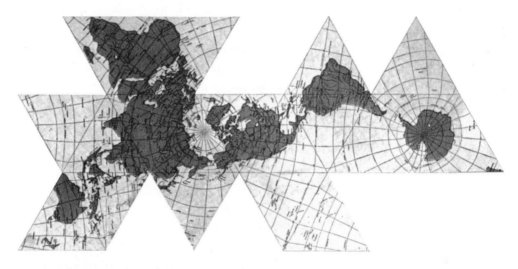

6. To begin your study of nets, consider possible nets for a cube.

   a. Examine the three nets shown below. Which of these nets can be folded to make a cube?

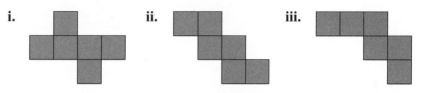

   b. Make a new net of your own that can be folded to make a cube.

   c. Which die could be formed from the net on the left?

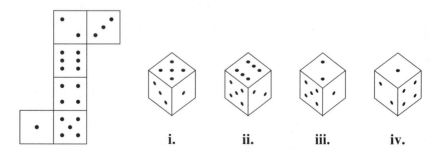

7. The equilateral triangle is more versatile than the square when it comes to making nets for space-shapes. You can make three different space-shapes using nets of only equilateral triangles attached to each other. Use copies of congruent equilateral triangles to complete this activity.

a. Attach four equilateral triangles edge-to-edge. What space-shapes can you form by folding your pattern? What *complete* (with no open faces) space-shapes can you form? Sketch the net and the complete space-shape formed. Find another pattern of the four triangles that can be folded into the same shape. Sketch your net. You have made a **regular tetrahedron**.

b. Now try your hand at making a space-shape with eight equilateral triangular faces. Draw a net of eight equilateral triangles that will fold into a space-shape. Sketch the net and the space-shape. You have made a **regular octahedron**.

c. Examine the net of a "globe" shown on the previous page.

- How many equilateral triangles are in this net?

- When the net is folded to form the "globe," how many triangles are at each vertex?

- Fuller's globe shown at the right is an example of a space-shape called a **regular icosahedron**. Make an icosahedron model.

Checkpoint

Nets can provide an efficient way of forming space-shapes whose faces are polygons.

ⓐ Sketch a net of squares that can be folded into an open-top box.

ⓑ How could you modify your net in Part a so that it could be folded into a model of a house with a peaked roof?

ⓒ Compare the sum of angle measures at a common vertex of a tiling with those of a net.

Be prepared to share your sketches, findings, and reasoning with the class.

▶ On Your Own

A cereal box is shown at the right. Draw a net for a model of it. Cut out your net and verify that it can be folded into a model of the box.

Modeling • Organizing • Reflecting • Extending

Modeling

1. Examine the picture at the right of a Native American rug.

   a. Are there any designs in this rug that have rotational symmetry? Sketch each design and describe the angles through which it can be turned.

   b. Are there any designs in this rug that have line symmetry? Sketch each design and the lines of symmetry.

   c. Are there any designs which have both rotational and line symmetry? If so, identify them. Where is the center of rotation in relation to the lines of symmetry?

2. Polygons and symmetry are important components of the arts and crafts of many cultures.

   a. The design of the quilt below is called "Star of Bethlehem." What rotational symmetry do you see in the fundamental "stars"? List the degree measure of each turn that will rotate each shape onto itself.

   b. What line symmetry do you see in the "stars"? Sketch to illustrate.

   c. Does the quilt as a whole have rotational or line symmetry? Describe each symmetry you find.

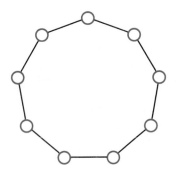

3. Here is a two-person game that can be played on any regular polygon. To play, place a penny on each vertex of the polygon. Take turns removing one or two pennies from adjacent vertices. The player who picks up the last coin is the winner.

   a. Suppose the game is played on a nine-sided polygon, as shown at the left. Try to find a strategy using symmetry that will permit the second player to win always. Write a description of your strategy.

   b. Will the strategy you found work if the game is played on any polygon with an odd number of vertices? Explain your reasoning.

   c. Suppose the game is played on a polygon with an even number of vertices, say an octagon. Try to find a strategy that will guarantee that the second player still can win always. Write a description of this strategy.

4. Objects in nature are often symmetric in form.

   a. The shapes below are single-celled sea plants called *diatoms*.

   ■ Identify all of the symmetries of these diatoms.

   ■ For those with reflection symmetry, sketch the shape and show the lines of symmetry.

   ■ For those with rotational symmetry, describe the angles of rotation.

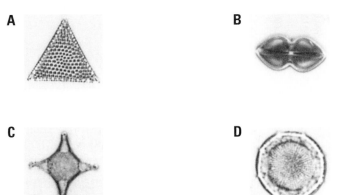

A   B

C   D

   b. Identify all of the symmetries of the two flowers shown below.

   ■ If the flower has line symmetry, sketch the shape and draw the lines of symmetry.

   ■ If the flower has rotational symmetry, describe the angles of rotation.

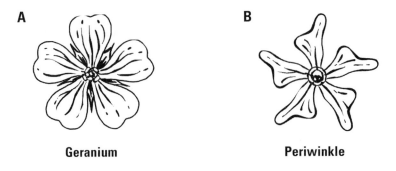

A   B

Geranium   Periwinkle

c. It has been said that no two snowflakes are identical.

■ Identify the symmetries of the snowflakes below.

■ In terms of their symmetry, how are the snowflakes alike?

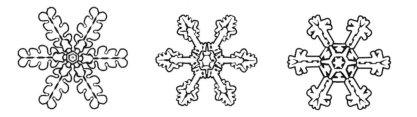

5. Great Lakes Packaging manufactures boxes for many different companies. Shown below is the net for one type of box manufactured for a candy company.

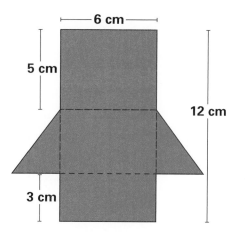

a. Sketch the box.

b. Sketch two other possible nets that could be used to manufacture the same box.

c. Find the volume of the box.

d. Find the surface area of the box.

e. Does the box have any symmetries? If so, explain how the symmetries are related to the symmetries of its faces.

Organizing

1. In Investigation 1 of this lesson, Ellen invented the rule $A = \frac{(n-2)180}{n}$ to predict the measure A of one angle of a regular n-gon.

a. Do you think Ellen's rule is correct? Explain your reasoning.

b. As the number of sides of a regular polygon increases, how does the measure of one of its angles change? Is the rate of change constant? Explain.

c. Use Ellen's rule to predict the measure of one angle of a regular 20-gon. Could a tessellation be made of regular 20-gons? Explain your reasoning.

d. When will the measure of each angle of a regular polygon be a whole number?

e. Use your calculator or computer software to produce a table of values for angle measures of various regular polygons. Use your table to help explain why only regular polygons of 3, 4, or 6 sides will tile a plane.

2. Examine the two histograms below.

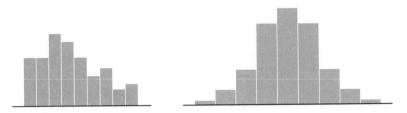

a. Locate any lines of symmetry for the two histograms.

b. On each histogram, estimate the location of the median and mean.

c. If a distribution is symmetric, what can you conclude about its median and mean? Explain your reasoning.

d. If a distribution is symmetric, what, if anything, can you conclude about its mode? Explain your reasoning.

3. Shown below are graphs of various relations between variables x and y. The scale on the axes is 1.

a. For each graph, locate any line of symmetry. Write the equation of the symmetry line.

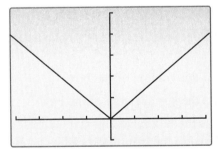 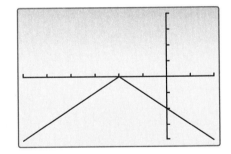

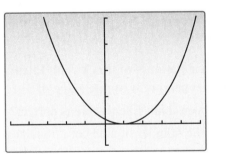 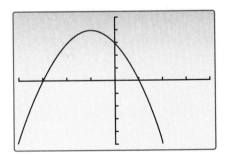

b. Suppose you have a graph and its line of symmetry is the *y*-axis. If one point on the graph has coordinates (–8, –23), what is the *y*-coordinate of the point on the graph with *x*-coordinate 8? Explain your reasoning.

4. In general, how many lines of symmetry does a regular *n*-gon have? How many rotational symmetries?

5. Circles have both line and rotational symmetries.

   a. Describe all the line and rotational symmetries of a circle as completely as possible.

   b. If you are given a circle, how can you find its center? Describe as many different ways as you can.

   c. How can you use a method for finding the center of a circle to help you find the center of rotation for a shape that has rotational symmetry?

6. The following two figures come from a branch of mathematics called *fractal geometry*. Describe all the line and rotational symmetries of each figure as completely as you can.

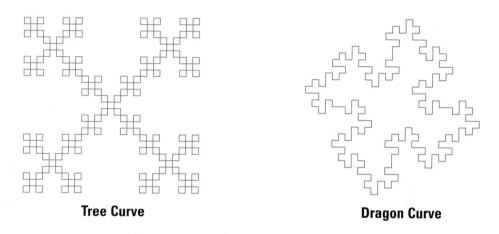

Tree Curve   **Dragon Curve**

Reflecting

1. Thumb through the yellow pages of a phone directory noting the shape of company logos. Why do you think so many of the logo designs are symmetric?

2. Do you find it easier to recognize line symmetry or rotational symmetry in a shape? What do you think might explain this fact?

3. Cross-cultural studies suggest that symmetry is a fundamental idea that all people use to help understand, remember, compare, and reproduce forms. However, symmetry preferences have been found across cultures. One study found that symmetry about a vertical line was easier to recognize than symmetry about a horizontal line. The study also found that symmetry about a diagonal line was the most difficult to detect. (Palmer, S.E. and K. Henenway. 1978. Orientation and symmetry: effects of multiple, rotational, and near symmetries. *Journal of Experimental Psychology* 4[4]: 691–702.)

   a. Would the findings of the study apply to the way in which you perceive line symmetry?

   b. Describe a simple experiment that you could conduct to test these findings.

4. Tiling patterns often are found on floors and walls. What tiling patterns are most common in homes, schools, and shopping malls? What might explain this?

5. In Investigation 2 of this lesson, you examined shapes that tessellate a plane—a flat surface. Examine at least three of the following balls: baseball, softball, basketball, tennis ball, volleyball, and soccer ball.

   a. Is it possible to tessellate a sphere?

   b. If so, describe the shapes of the tiles.

   c. Find a National Geographic magazine with a world map in it. How do the map-makers use the idea of tessellations in their production of world maps reproduced on a flat surface?

Extending

1. Using a computer drawing utility or a straightedge and compass, design a simple quilt pattern. Try to use polygonal shapes which were not used in the designs shown in this lesson. Print or draw your pattern and then indicate all symmetry lines of the design.

2. How could you use a circle to draw a 10-sided polygon with sides of equal length? Try it. Does the polygon have rotational symmetry? If so, through what angles can it be turned? Where is the center?

3. Find as many nets of six squares as you can that will fold to make a cube. Sketch each. (*Hint:* There are more than ten!)

4. The space-shape shown here is made of 12 regular pentagons. It is called a **regular dodecahedron**.

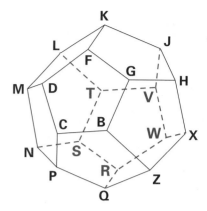

    a. Use a net of pentagons to make a dodecahedron shell.

    b. Make a die by numbering the faces.

    c. Toss the die 100 times. Record the number of times each face lands up.

    d. Make a histogram of your data in Part c. What do you observe?

    e. Repeat Parts a–d for a dodecahedral die for which the opposite faces add to 13. Compare your histograms.

5. The **diagonals** of a polygon are segments connecting pairs of vertices which are not endpoints of the same side.

    a. Examine the diagonals of each type of quadrilateral. For which quadrilaterals are the diagonals lines of symmetry?

    b. Develop a formula for the number of diagonals in any *n*-gon.

6. A *kapa pohopoho* is a Hawaiian quilt made from twelve or more unique designs. The original designs exhibit reflection and rotational symmetry. Each design is cut from one piece of fabric and sewn onto a square piece of background fabric. The following steps illustrate one way to create a Hawaiian quilt design.

"Kualoa" © Helen M. Friend, 1991.

    a. Fold a piece of square paper in half by bringing the bottom side up to meet the top. Fold this half portion into a square. Note which corner of the new square is the center of the original square. Fold along the diagonal that has this corner as one vertex. You now should have a right triangle with one end of the hypotenuse at the center of the original square. The other end is at the point where the four corners of the original square meet. Sketch a design along the leg of the right triangle adjacent to the four corners of the original square. Cut along your design.

    b. Unfold your pattern. If the design has reflection symmetry, sketch in the lines of symmetry. If the design has rotational symmetry, identify the center and angles of rotation.

    c. If you open up the folded square before making any cuts, what lines should your folds represent? (You can repeat Part a to check.)

    d. How does your design compare with those of your classmates?

    e. What would happen if you started with a circle or an equilateral triangle instead of a square? Can you make designs with only one line of symmetry? With two lines? With more than two lines? Explain your reasoning.

INVESTIGATION 3 ▶ Symmetry Patterns in Strips

Many plane-shapes and space-shapes have symmetry, either about a line, a point, or a plane. Artists have used these types of symmetry for centuries. For example, from earliest history, space-shapes often have been decorated with *strip patterns*.

1. Imagine slicing the Native American jar at the right by a symmetry plane and then "flattening out" the shape.

   a. Draw what you think would be the strip pattern along the center of the plane-shape.

   b. What is it about the pattern that permitted you to draw the full pattern without seeing the back of the jar?

2. Shown below are some general strip patterns commonly used for decorative art. Imagine that each strip pattern extends indefinitely to the right and to the left.

   a. Examine each pattern and make a sketch of the next two shapes to the right in the pattern.

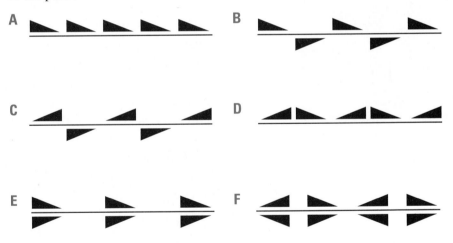

   b. Examine the design symmetries of the two bracelet patterns below. Match each pattern with the strip pattern in Part a which it most resembles.

c. Each of the strip patterns in Parts a and b exhibit slide or *translational symmetry*. What are two essential characteristics of the patterns that ensure translational symmetry?

d. Write a definition of translational symmetry for strip patterns. Compare your group's definition with that of other groups. Work out any differences.

3. The portions of the strip patterns below come from artwork on the pottery of San Ildefonso Pueblo, New Mexico.

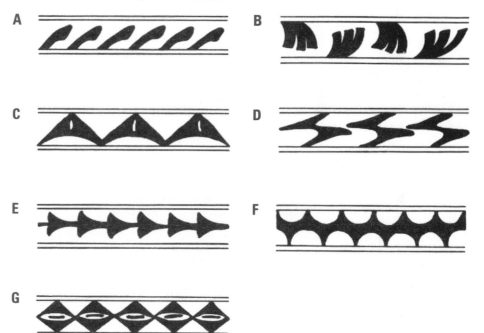

Source: Crowe, D.W. and D.K. Washburn. "Groups and Geometry in the Ceramic Art of San Ildefonso." *Algebras, Groups and Geometries* 2, no. 3 (September 1985): 263–277.

a. Confirm that each pattern has translational symmetry.

b. Examine each strip for reflection and rotational symmetries. If a strip has such a symmetry, it appears the same before and after it is reflected across a line or rotated about a point. Systematically record your results.

c. Look again at strip pattern B. It has neither rotational nor reflection symmetry. However, consider the following *transformation*:

Slide the strip to the right so the first "shape" is above the second "shape." Now reflect the entire strip over the horizontal line through the middle of the strip.

Does the pattern of the strip transform onto itself?

d. Make a sketch of your footprints as they would appear if you walked a straight line in snow or damp sand. How is your footprint strip pattern related to the transformation described in Part c?

e. Using the shape below, draw a strip pattern that has the type of symmetry described in Part c.

The transformation described in Part c of Activity 3 is called a **glide reflection** because you glide (translate) and reflect across a line parallel to the direction of the glide.

4. Re-examine the seven strip designs of the San Ildefonso Pueblo on page 403. Do any of these patterns have glide reflection symmetry? If so, record it with your other results from Part b of Activity 3.

Checkpoint

Strip patterns by design have translational symmetry. But often they have other symmetries as well.

ⓐ What symmetries, other than translational symmetry, are evident in some strip patterns?

ⓑ Describe how you can test a strip pattern for various symmetries.

Be prepared to share and explain your group's descriptions.

▶ **On Your Own**

Design an interesting strip pattern that has reflection symmetry about a horizontal line and has translational symmetry.

INVESTIGATION ▶ 4 Symmetry Patterns in Planes

Strip patterns often form a decorative border for other designs that cover part of a plane. Plane tiling patterns are far more numerous than strip designs. They can be found in the floor coverings, ceramic tile work, and textiles of many cultures. The walls and floors of the Alhambra, a thirteenth-century Moorish palace in Granada, Spain, contain some of the finest early examples of this kind of mathematical art. Note the variety of their patterns in the photo at the top of the following page.

The Alhambra in Granada, Spain

Recently, the tessellation artwork of the Dutch artist, M.C. Escher, has become very popular. Escher was deeply influenced by the work of the ancient Moors. Examine the following Escher pattern.

1. As a class, imagine that this pattern is extended indefinitely to cover the plane.

   a. Describe all translational symmetries that you see.

   b. Does the pattern have reflection or rotational symmetries? If so, describe them.

   c. Does the pattern have glide reflection symmetry? If so, describe it.

   d. On what type of polygon do you think this pattern is based?

Tessellations such as that above are based on polygon tilings of the plane. The polygon is modified carefully so that the new shape will tile the plane when certain transformations are applied. Rotations and translations are two of the most common transformations. Escher was a master at these modifications.

Knowing which polygonal regions tessellate a plane is one important part of understanding Escher-like tilings. Another important aspect is understanding which transformations will take an individual tile into another tile within the entire tiling pattern. Applying these ideas leads to beautiful patterns.

2. A tiling based on squares is shown in the figure below. Adjacent squares have different colors. Think of the square labeled "0" as the beginning tile. How many squares surround square 0?

	1	2	3		
	8	0	4		
	7	6	5		

a. Which transformations (reflection, rotation, translation, or glide reflection) will move square 0 to a surrounding square of the *same* color? Illustrate with diagrams.

b. Which transformations will move square 0 to a surrounding square of the *other* color? Illustrate with diagrams.

c. If *translation* is the only motion permitted, can square 0 be translated to each of squares 1–8? On a copy of the tiling, represent each such translation by drawing an arrow which shows its direction and how far square 0 must be translated.

d. Using only *rotation* about a vertex of square 0, can square 0 be moved onto each of squares 1–8? Describe each such rotation on the copy of the tiling by marking its center and giving its angle. If some squares can be reached in more than one way, describe each way.

Escher used the square tiling and knowledge of the translations that move square 0 to the adjacent positions to help him design some of his art. In the remainder of this investigation, you will explore some of the mathematics behind Escher's art.

3. If you were to replace square 0 with the shape at the right, could you make a tiling? If so, sketch the resulting pattern on a sheet of paper.

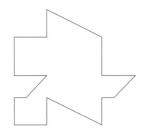

  a. The shape you used is a modification of square 0. Describe as accurately as you can how you think square 0 was modified to obtain this shape.

  b. How do the areas of square 0 and the modified shape compare?

The above shape, like most of Escher's shapes, is an example of an **asymmetric** shape that tiles the plane. It was created by modifying a symmetric shape, the square, that also tiled. This technique can be used to create an unlimited number of interesting tessellations.

4. Working alone, modify the sides of square 0 to create a shape that is asymmetric and that will tile the plane by translation to the eight adjacent positions. Sketch or make a computer-generated nine-square version of your tessellation. Compare your tessellation with those of the other members of your group.

Checkpoint

In this investigation, you explored how a symmetric shape that tessellates the plane can be transformed to an asymmetric shape that also tessellates.

ⓐ Suppose that by translation alone, you have created a tessellation of an asymmetric shape based on a square.

  ■ How must the "sides" of the shape be related?

  ■ How are the areas of the original square and the transformed asymmetric shape related? Why?

ⓑ Describe an algorithm for transforming any parallelogram shape into an Escher-like shape that will tile the plane.

Be prepared to discuss your group's ideas with the class.

▶ On Your Own

Recall that a rhombus is a parallelogram, all of whose sides are the same length. Can a rhombus be made into an Escher-like asymmetrical tile that will tessellate by translation alone? Explain and illustrate your answer.

Modeling • Organizing • Reflecting • Extending

Modeling

1. Archeologists and anthropologists use symmetry of designs to study human cultures. The table below gives frequency of strip designs on a sampling of pottery from two cultures on two different continents: the Mesa Verde (U.S.A.) and the Begho (Ghana, Africa).

Strip Pattern Symmetry Type	Mesa Verde		Begho	
	Number of Examples	Percentage of Total	Number of Examples	Percentage of Total
Translation Symmetry Only	7		4	
Horizontal Line Symmetry	5		9	
Vertical Line Symmetry	12		22	
180° Rotational Symmetry	93		19	
Glide Reflection Symmetry	11		2	
Glide Reflection and Vertical Line Symmetry	27		9	
Both Horizontal and Vertical Line Symmetry	19		165	
Total	174		230	

Source: Washburn, Dorothy K. and Donald W. Crowe. *Symmetries of Culture: Theory and Practice of Plane Pattern Analysis.* Seattle: University of Washington Press, 1988.

a. Copy the chart and fill in the two "Percentage of Total" columns.

b. Which types of symmetry patterns appear to be preferred by the Mesa Verde? By the Begho?

c. Examine the strips shown below. In which place is each strip more likely to have been found? Use the data from the table to explain your answer.

i.   ii.   iii.

2. Examine the eighteenth-century embroidered-cloth strip patterns of Kuba, Democratic Republic of Congo, shown below.

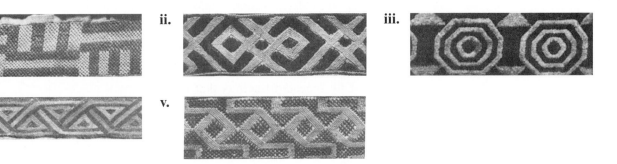

i. ii. iii.

iv. v.

a. Identify those patterns that exhibit reflection symmetry about a vertical line.

b. Identify those patterns that exhibit reflection symmetry about a horizontal line.

c. Identify those patterns that exhibit rotational symmetry.

d. Identify those patterns that exhibit glide reflection symmetry.

3. The 14 Japanese border designs pictured below seem to have great variability. Sort them into groups. Describe the characteristics used to determine membership in the groups you create.

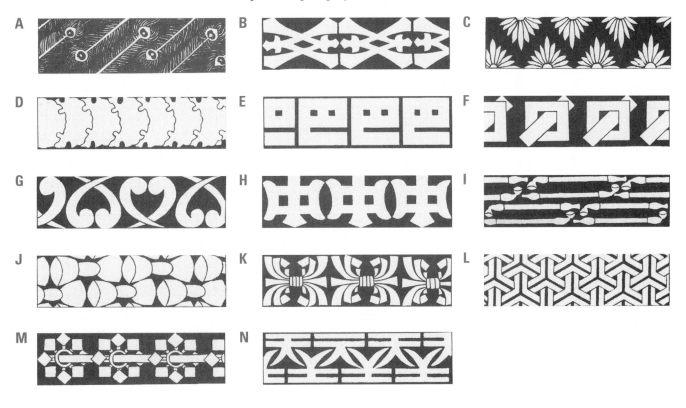

4. Shown here is an example of a common first step used to create a tessellation. This example could become the "flying-horse" tessellation shown on page 406.

    a. Trace the modified square. Complete the modification to obtain the horse shape.

    b. Test by repeated tracings that the shape will tessellate. If necessary, make further adjustments for a good fit.

    c. What kinds of symmetry does your tessellation have?

5. Create an Escher-like tessellation of your own by modifying a square so that it will tessellate the plane by translation alone. Try to construct your shape so that it can be enhanced to look like a common object or animal. (You may use both curved and straight segments.)

Organizing

1. For each of the following descriptions, construct a strip pattern that has translational symmetry and the given symmetry or symmetries. Use a computer drawing utility if it is available.

    a. vertical line symmetry   **b.** horizontal line symmetry

    c. 180° rotational symmetry   **d.** horizontal and vertical line symmetry

    e. glide reflection symmetry   **f.** glide reflection and vertical line symmetry

2. Use the chart below to help organize your thinking about classifying strip patterns.

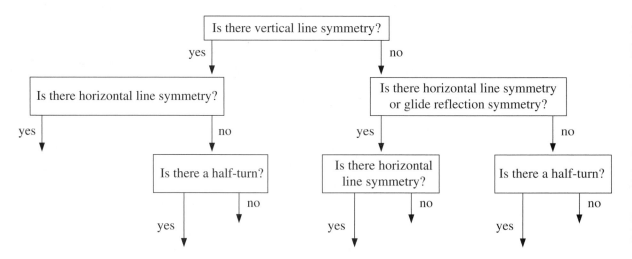

Source: Washburn, D. and D. Crowe. *Symmetries of Culture.* Seattle: University of Washington Press, 1988.

a. Make or obtain an enlarged copy of this chart. For each of the triangle strip patterns in Activity 2 of Investigation 3 (page 402), sketch the pattern at the point on the chart that describes its characteristics.

b. The chart suggests that there are seven different possible one-color strip patterns. Which pattern was not included in Activity 2? Draw an example of the missing strip pattern at the appropriate place on the chart.

3. Suppose the following graphs extend indefinitely in the pattern illustrated. Which have translational symmetry? For those that do, identify any other symmetries of the graph.

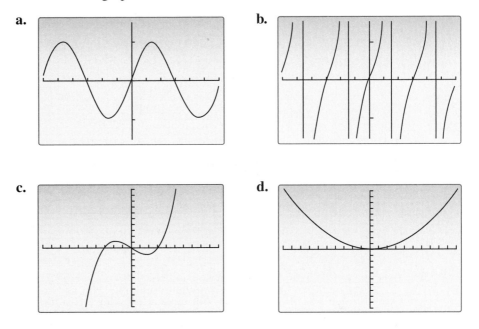

a.

b.

c.

d.

4. Suppose each polygon below is modified as shown.

a. Explain why you think the shape will or will not tessellate the plane.

b. If the shape will tessellate, describe the symmetries, if any, of the resulting tessellation.

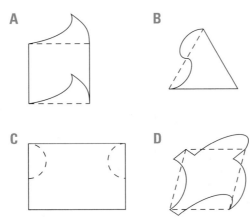

A

B

C

D

Reflecting

1. Juan has been trying to make a strip pattern which has both horizontal and vertical symmetry, but not 180° rotational symmetry. He asks your advice. What do you suggest and why?

2. The chart in Organizing Task 2 provides a systematic way of analyzing strip patterns. On the basis of the chart, why can you conclude there are exactly seven essentially different strip patterns?

3. In your home or neighborhood, find examples of strip patterns. Investigate why the patterns were chosen or if the patterns have special meaning. Make a sketch of each design. Describe the symmetries evident in the designs.

4. Dorothy Washburn is an archeologist. She discovered that the pattern found in Escher's tessellation of lizards is strikingly similar to that found in a Fiji basket lid and an Egyptian wall mosaic. Why do you think patterns of this sort are found in different cultures?

5. In 1970, M.C. Escher wrote: "Although I am absolutely innocent of training or knowledge in the exact sciences, I often seem to have more in common with mathematicians than with my fellow artists." (Escher, M.C. *The Graphic Work of M.C. Escher*, trans. John E. Brigham. New York: Ballantine Books, 1971.)

   a. What is it about mathematics, geometry in particular, that permits people with little mathematical training to discover, on their own, many of its basic principles?

   b. Based on what you have seen of Escher's work, why do you think he felt as he did about himself?

Extending

1. In your previous study of mathematics, you may have observed that decimal representations of numbers often have interesting patterns.

   a. Investigate how the decimal representation of numbers between 0 and 1 might be related to strip patterns. Identify any characteristics of the numbers that make the analogy fail.

   | 1/11 | .0909090909 |
   | π/4 | .7853981634 |

   b. What would be the fundamental unit for the strip pattern for $\frac{2}{11}$? For $\frac{1}{13}$? For $\frac{1}{3}$?

   c. What would you say about the strip pattern representation of $\frac{1}{2}$?

2. The strip patterns you have investigated are often classified using symbols. The following one seems useful. It is made up of 3 characters in order. The first character is **m** if there is a vertical reflection and **1** (one) if there is none; the second character is **m** if there is a horizontal reflection, **a** if there is a glide reflection, and **1** if there is neither; the third character is **2** if there is 180° rotational symmetry and **1** otherwise.

a. Assign a symbol to each of the 7 San Ildefonso Pueblo designs studied in Activity 3 of Investigation 3 (page 403).

b. Assign a symbol to each of the Japanese border patterns in Task 3 of the Modeling section.

c. Assign a symbol to each of the strip patterns shown below.

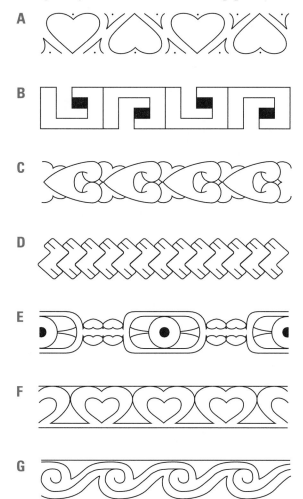

3. Shown below are two sketches that M.C. Escher used in his lectures about ways to regularly fill or divide the plane. Each of these tilings is based on a square. Where are the vertices of the squares located?

  a. In the sketch at the left, what transformations will move a tile of one color onto a tile of the same color?

  b. What transformations will move a tile of one color onto a tile of the other color? How has Escher indicated this?

  c. Make your own tile with characteristics similar to those found in this sketch. Use it to make a tiling of the plane.

verschuiving en assen.

verschuiving en glijspiegeling.

  d. In the sketch at the right above, what transformations will move a tile of one color onto a tile of the same color?

  e. What transformations will move a tile of one color onto a tile of the other color? How has Escher indicated this?

  f. Make a personal tile with characteristics similar to those found in this sketch. Use it to make a tiling of the plane.

4. The Escher tessellation at the left below is based on a hexagon.

    a. What characteristic of the tessellation helps you locate each vertex of the hexagon?

    b. Draw a regular hexagon on poster board. Modify it to make an Escher-like shape. Describe the modifications which you made. Show by repeated tracings of the shape that it will cover the plane.

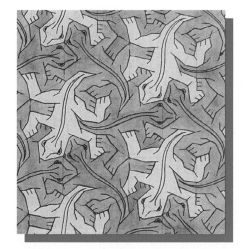 

5. Study the tessellation of light and dark-colored lizards at the right above.

    a. What is the fundamental polygonal shape on which this tessellation is based? Justify your choice and discuss it with a classmate.

    b. Make a poster board model of the polygon on which the tessellation is based. Show the modifications needed to make the shape used in the design.

    c. Compare the areas of the polygon and the lizard tiles.

    d. What transformation will move a light-colored lizard with its head facing upward to:

       ■ a light-colored lizard with its head pointing down?

       ■ a dark-colored lizard with its head toward the right?

       ■ a dark-colored lizard with its head toward the left?

Looking Back

In this unit, you saw how space-shapes and plane-shapes are related. You also saw how they are constructed and visualized, and how they may be drawn in two dimensions. You learned ways to measure them and what sorts of symmetry they have. You even saw how they may be used in art and design to make our lives more productive and enjoyable, both visually and physically. The landscape architects who designed this mall garden area made extensive and integrated use of these fundamental ideas of geometry.

This final lesson of the unit gives you the opportunity to pull together and apply what you have learned. You will use your visualization skills and knowledge of shapes, symmetry, tessellations, and measurement in new situations.

1. Shown below is a net for a decorative box. Some of its dimensions are given.

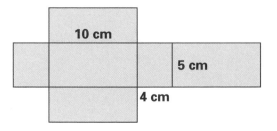

a. Name the space-shape for which this is a net.

b. Make a paper model of the space-shape.

c. Sketch the space-shape showing its hidden edges. Give the lengths of each edge.

d. Describe any planes of symmetry for the space-shape.

e. Draw and name each face of the space-shape.

f. Describe all symmetries for each face.

g. Find the surface area and the volume of the space-shape. How is the surface area related to the area of the net?

h. Will a pen 11 cm long fit in the bottom of the box? Explain why or why not in at least two different ways.

i. Estimate the length of the longest pencil that will fit inside the box. Illustrate and explain how you found your answer.

j. Draw a different net for the same space-shape.

2. As an art project, Tmeeka decided to make a decorative baby quilt for her newborn sister Kenya. She chose the quadrilateral below and modified it as shown. Her completed pattern is shown next to the fundamental tile.

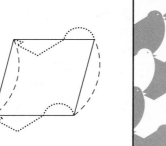
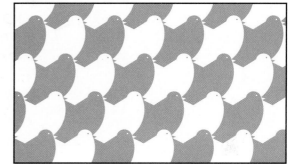

a. Does this quadrilateral have a special name? If so, give it; if not, explain.

b. Describe how Tmeeka modified the sides of the quadrilateral to make the fundamental tile. What transformations did she use?

c. Use the geometric ideas and language developed in this unit to describe how Tmeeka made the quilt pattern.

d. Assuming the pattern continues in all directions, describe all the symmetries you can find.

e. Consider only the bottom row of shapes. Does this row form a strip pattern? If so, describe its symmetries including any centers of rotation and any lines of reflection.

f. Create a strip pattern with both horizontal and vertical line symmetry that could be used as a border for Tmeeka's quilt. Describe all symmetries in your strip pattern.

Checkpoint

In this unit, you have developed ways for making sense of situations involving shapes and spatial relationships.

ⓐ Compare and contrast prisms, pyramids, cylinders, and cones.

ⓑ Compare and contrast the six special quadrilaterals.

ⓒ For each shape in Parts a and b, describe a real-life application of that shape. What properties of the shape contribute to its usefulness?

ⓓ Describe a variety of ways you can represent (draw or construct) space-shapes.

ⓔ Pose a problem situation which requires use of cubic, square, and linear units of measure. Describe how you would solve it.

ⓕ Describe how to make a strip pattern and how to make a tiling of the plane.

Be prepared to share your descriptions with the entire class.

▶On Your Own

Write, in outline form, a summary of the important mathematical concepts and methods developed in this unit. Organize your summary so that it can be used as a quick reference in future units and courses.

Exponential Models

Unit **6**

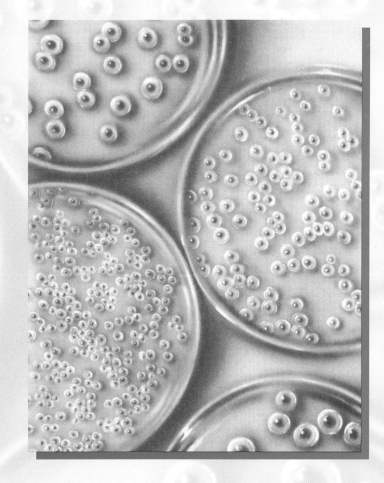

Lesson 1 ▸ *Exponential Growth*

News stories spread rapidly in modern society. With broadcasts over television and radio, millions of people hear about important events within hours. The major television and radio news networks try hard to report only stories that they know are true. But quite often, rumors get started and spread around a community by word of mouth alone.

Suppose that to study the spread of information through rumors, two students started this rumor at 5 PM one evening:

Because of the threat of a huge snowstorm, there will be no school tomorrow and probably for the rest of the week.

The next day, they surveyed students at the school to find out how many heard the rumor and when they heard it. How fast do you think this rumor would spread?

Think About This Situation

The graphs below show three possible patterns in the rate at which the school-closing rumor could spread.

a How would you describe the rate of rumor-spread in the case of each graph?

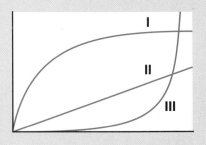

b Which pattern of spread is most likely if the students plant the story on the 5 o'clock television or radio news? Explain your reasoning.

c Which pattern of spread is most likely if the rumor spreads only by word of mouth around the community? Why?

In many problems, key variables are related by linear models. But there are many other important situations in which variables are related by nonlinear patterns. Some examples include the spread of information and disease, changes in populations, pollution, temperature, bank savings, drugs in the bloodstream, and radioactivity. These situations often require mathematical models with graphs that are curves. Equations for the models use forms other than the familiar $y = a + bx$. In this unit, you will learn to use the family of nonlinear models that describes *exponential* patterns of change.

INVESTIGATION 1 ▶ Have You Heard About … ?

Some organizations need to spread accurate information to many people quickly. One way to do this efficiently is to use a telephone calling tree. For example:

> The Silver Spring Soccer Club has boys and girls from about 750 families who play soccer each Saturday in the fall. When it is rainy, everyone wants to know if the games are canceled. The club president makes a decision and then calls two families. Each of them calls two more families. Each of those families calls two more families, and so on.

This calling pattern can be represented by a **tree graph** that starts like this:

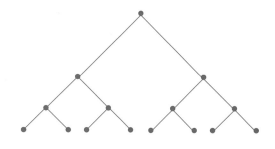

1. What do the vertices of this tree graph represent? What do the edges represent?

2. At the start of the calling process, only the president knows whether the games are on or not. In the first stage of calling, two new families get the word. In the next stage, four others hear the decision, and so on.

   a. Make a table showing the number of families who will hear the decision at each of the next eight stages of the calling process. Then plot the data.

Stage of Calling Tree	0	1	2	3	4	5	6	7	8	9	10
Families Informed	1	2	4								

b. How does the number of families hearing the message grow as the calling tree progresses in stages? How is that pattern of change shown in the plot of the data?

c. How many stages of the tree will be needed before all 750 families know the decision? How many telephone calls will be required?

3. How will word pass through the club if each person in the tree calls three other families, instead of just two?

a. Make a tree graph for several stages of this calling plan.

b. Make a table showing the number of families who will hear the decision at each of the first ten stages of the calling process. Then plot the data.

Stage of Calling Tree	0	1	2	3	4	5	6	7	8	9	10
Families Informed		1	3								

c. How does the number of families hearing the message increase as the calling tree progresses in stages? How is that pattern of change shown in the plot of the data?

d. How many stages of the tree will be needed before all 750 families know the decision? How many telephone calls will be required?

4. In each of the two calling trees, you can use the number of phone calls at any stage to calculate the number of calls at the next stage.

a. Use the words *NOW* and *NEXT* to write equations showing the two patterns.

b. Explain how the equations match the patterns of change in the tables of (*stage, number of families informed*) data.

c. Describe how the equations can be used with your calculator or computer to produce the tables you made in Activities 2 and 3.

d. Write an equation relating *NOW* and *NEXT* that could be used to model a telephone calling tree in which each family calls four other families.

Look back at the patterns of change in number of families informed by the two calling trees.

ⓐ Compare the calling trees by noting similarities and differences in the following:

  ■ patterns of change in the tables of (*stage, number of families informed*) data;

  ■ patterns in the graphs of (*stage, number of families informed*) data;

  ■ the equations relating *NOW* and *NEXT* numbers of calls.

ⓑ Below are a table and a graph for a linear model. In what ways are the table, graph, and equation patterns for the calling trees different from those of linear models?

x	y
0	1
1	2
2	3
3	4
4	5

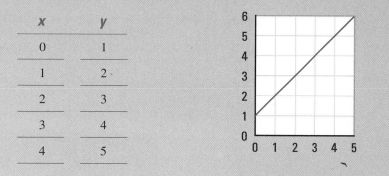

Be prepared to share your group's ideas with the rest of the class.

On Your Own

The patterns of change in information spread by calling trees occur in many other situations. For example, when bacteria infect some part of your body, they grow and split into two genetically equivalent cells again and again.

a. Suppose a single bacterium lands in a cut on your hand. It begins spreading an infection by growing and splitting into two bacteria every 20 minutes.

  ■ Make a table showing the number of bacteria after each 20-minute period in the first three hours. (Assume none of the bacteria are killed by white blood cells.)

- Plot the (*number of 20-minute time periods, bacteria count*) data.
- Describe the pattern of growth of bacteria causing the infection.

b. Use *NOW* and *NEXT* to write an equation relating the number of bacteria at one time to the number 20 minutes later. Then use the equation to find the number of bacteria after fifteen 20-minute periods.

c. How are the table, graph, and equation of bacteria growth similar to, and different from, the calling tree examples? How are they similar to, and different from, typical patterns of linear models?

INVESTIGATION ▶ 2 Shortcut Calculations

Everyone knows that mathematics is useful in solving business, engineering, or science problems. It is also used to design works of music and art. Sometimes it even plays a role in the plots of stories and books. For example, an old Persian legend illustrates the speed of exponential growth.

A wealthy king was rescued from danger by the quick thinking and brave action of one of his soldiers. The king wanted to honor the poor soldier, so he offered a very generous reward: a beautiful chessboard made of ivory and ebony and a set of gold chess pieces.

While the chess set was beautiful and valuable, the young man asked for a different reward. To help the poor people in his country, he asked the king to distribute rice from his storehouse—two grains for the first square of the chessboard, four grains for the second square, eight grains for the third square, sixteen grains for the fourth square, and so on. The king was pleased that he could keep his beautiful chessboard and repay the brave soldier with such a simple grant of rice to the poor. But he soon discovered that the request was not as simple as he thought.

1. Use your calculator to find the number of grains of rice for each of the squares 10, 20, 30, 40, 50, 60, and 64.

2. The national debt of the United States in 2000 was about $5,600,000,000,000. How does this number compare to the number of grains of rice for square 64 of the king's chessboard?

3. For some kinds of rice it takes about 2,000 grains to fill one cubic inch of space. Consider the number of grains that the king owed for the 64th square alone.

   a. How many cubic inches would that rice occupy?

   b. How many cubic feet?

   c. How many cubic yards?

   d. How many cubic miles? (There are 5,280 feet in 1 mile.)

4. A cubic inch of rice costs about $0.05. What would this mean for the present-day value of the rice on square 64 alone?

5. To calculate the number of grains of rice for each square of the king's chessboard, you could use the equation $NEXT = NOW \times 2$, beginning at 2 grains for the first square. So the number of grains of rice for square 2 could be represented as 2×2.

   a. Why can the number of grains of rice for square 3 be expressed as $(2 \times 2) \times 2$?

   b. Write expressions for the number of grains of rice for squares 5, 10, and 20.

   c. What is the shorthand way of writing the calculations in Part b using *exponents*?

   d. Write an exponential expression for the number of grains of rice for square 64. For any square x.

   e. Compare your exponential expressions in Parts c and d with those of another group. Resolve any differences.

6. You can use your graphing calculator or computer software and the exponential rule for any square x to make tables and graphs of the pattern formed by counting rice grains. Enter the rule in the "Y=" list of your calculator or computer, using the $\boxed{\wedge}$ key before the exponent. (With some tools, you may need to use the y^x or a^b key or a different symbol instead.)

   a. Make a table showing the number of grains of rice for squares 1 through 10. You may use the calculator program *Tableplot* (TBLPLOT), if available. The program allows you to switch easily between the table and a scatterplot of the table's values.

b. Use TBLPLOT or similar software to plot the (*square number*, *number of grains of rice*) data.

c. Explain why the table and plot produced using your exponential rule are the same as those produced using the equation $NEXT = NOW \times 2$, starting at 2.

7. Suppose the wealthy Persian king offered his soldier a more generous deal: 3 grains of rice for the first square, 9 for the second, 27 for the third, 81 for the fourth, and so on.

a. Use an equation relating *NOW* and *NEXT* and your calculator or computer software to find the number of grains of rice for each of the first 10 squares of the chessboard in this case.

b. Write a rule using exponents that could be used to calculate the number of grains of rice for any square, without knowing the amount on the previous square.

c. Enter your rule for Part b in the "Y=" list of your graphing calculator or computer software. Find the number of grains of rice for squares 15, 25, and 35.

d. For which square will the number of grains of rice first exceed 1 billion?

Checkpoint

Look back at the patterns of change in the situation involving the king's chessboard and those in the bacterial growth and telephone calling tree situations in Investigation 1.

ⓐ How are the patterns of change in the tables for the king's chessboard similar to, and different from, those in the telephone trees and bacterial growth problems?

ⓑ How are the graphs of those relations similar to each other and how are they different?

ⓒ Compare the equations modeling the three situations.

■ How are the rules using *NOW* and *NEXT* similar and how are they different?

■ Write exponential rules ($y = \ldots$) that model the telephone trees. Write a rule that models the bacterial growth problem.

■ How are these rules similar and how are they different?

Be prepared to share your ideas with the entire class.

The patterns of change in the situations involving the king's chessboard, bacterial growth, and telephone calling trees are called **exponential growth**. Exponential growth patterns of change can be modeled using rules involving exponents.

On Your Own

The sketch below shows the first stages in the formation of a geometric figure. This figure is an example of a *fractal*. At each stage in the growth of the figure, the middle of every segment is replaced by a triangular tent. The new figure is made up of more, but shorter, segments.

Start

Stage 0   Stage 1   Stage 2

a. Make a sketch showing at least one more stage in the growth of this fractal. Describe any symmetries that the fractal has at *each* stage.

b. Continue the pattern begun in this table:

Stage of Growth	0	1	2	3	4	5	6	7
Segments in Design	1	4						

c. Write an equation showing how the number of segments at any stage of the fractal can be used to find the number of segments at the next stage.

d. Write an exponential rule that can be used to find the number of segments in the pattern at any stage *x*, without finding the numbers at each stage along the way. Begin your rule, "$y = \dots$."

e. Use the rule from Part d to produce a table and a graph showing the number of segments in the fractal pattern at each of the first 10 stages of growth. Do the same for the first 20 stages. (The calculator program TBLPLOT is helpful here.)

f. At what stage will the number of small segments first reach 1 million?

INVESTIGATION ▶ 3 Getting Started

Bacterial infections seldom start with a single bacterium. Suppose that you cut yourself on a rusty nail that puts 25 bacteria cells into the wound. Suppose also that those bacteria divide in two after every quarter of an hour.

1. Use *NOW* and *NEXT* to write an equation showing how the number of bacteria changes from one quarter-hour to the next.

2. Make a table showing the number of bacteria in the cut for each quarter-hour over the first three hours. Then plot a graph of the (*number of quarter-hours, bacteria count*) data.

3. In what ways are the table, graph, and equation of bacteria counts in this case similar to, and different from, the simple case (pages 423–424) that started from a single one-celled bacterium?

You could use the equation from Activity 1 to find the number of bacteria after 8 hours. (That would assume your body did not fight off the infection and you did not apply any medication.) Activity 4 will help you find a way to get that answer directly, without finding the bacteria count at each quarter-hour along the way.

4. Begin by making an estimate of the number of bacteria after 8 hours.

   a. What arithmetic operations are required to calculate the bacteria count after 8 hours (32 quarter-hours) if the equation relating *NOW* and *NEXT* is used?

   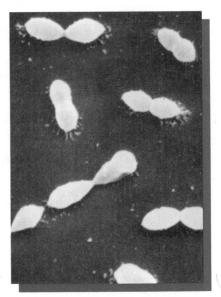

   - How can those operations be written in short form using exponents?

   - What set of calculator keystrokes will give the result quickly?

   - What is the number of bacteria after 8 hours? How close was your estimate?

   b. Write a rule using exponents that could help calculate the bacteria count. Then use your rule to calculate the number of bacteria after 8 hours and compare your answer to that in Part a.

5. Investigate the number of bacteria expected after 8 hours if the starting number of bacteria is 30, 40, 60, or 100, instead of 25. For each starting number, do the following. (Divide the work among your group members.)

   a. Find the number of bacteria after 8 hours.

   b. Write two equations that model the bacterial growth. One should use *NOW* and *NEXT*. The other should begin "*y* = … ."

   c. Make a table and plot of (*number of quarter-hours, bacteria count*) data.

d. Compare your results.

  ■ How were your calculations of the number of bacteria after 8 hours similar and how were they different?

  ■ How are the equations relating *NOW* and *NEXT* and the equations beginning "*y* = …" for bacteria counts similar and how are they different?

  ■ How are the tables and graphs of (*number of quarter-hours, bacteria count*) data similar and how are they different?

Just as bacterial growth won't always start with a single cell, other exponential growth processes can start with different initial numbers. Think again about the telephone calling tree for the Silver Spring Soccer Club in Investigation 1.

6. Suppose that before deciding to call off play because of bad weather, the president must talk with the club's four-member board of directors. When the calling tree is started, there are already five people who know the news to be spread. The president and each member of the board begin the calling tree by calling three other families apiece. Each family then calls three other families, and so on.

  a. Draw a tree graph illustrating this pattern of calling.

  b. Write an equation showing how the number of calls at any stage of the calling process can be used to find the number of calls at the next stage.

  c. Use the equation from Part b to make a table and a plot showing how the number of calls increases as the process moves to further stages.

  d. What arithmetic operations are required to find the number of calls at the 8th stage of the tree, using the equation relating *NOW* and *NEXT*?

    ■ How can those operations be written in short form using exponents?

    ■ What set of calculator keystrokes will give the result quickly?

    ■ What is the number of calls at Stage 8?

  e. What rule using exponents could help with the number of call calculations?

7. Suppose the board of directors had only three members (so four people know the news at the start), and each caller in the tree is expected to call five other families.

  a. How would your answers to Activity 6 change?

  b. Which phone tree should reach all families in the least amount of time? Why?

In studying exponential growth, it is common to refer to the *starting point* of the pattern as **Stage 0** or the **initial value**.

8. Use your calculator and the $\boxed{\wedge}$ key to find each of the following values: 2^0, 3^0, 5^0, 23^0.

   a. What seems to be the calculator value for b^0, for any positive value of b?

   b. Recall the examples of exponential patterns in bacterial growth and telephone calling trees. How is your conclusion for Part a supported by these examples?

9. Now use your calculator to make tables of (x, y) values for each of the following equations. Use values for x from 0 to 10. Share the work among members of your group.

   i. $y = 5(2^x)$ **ii.** $y = 4(3^x)$

   iii. $y = 3(5^x)$ **iv.** $y = 7(23^x)$

   a. What patterns do you see in your tables that show how to model exponential growth from any starting point?

   b. If you see an equation of the form $y = a(b^x)$ relating two variables x and y, what will the values of a and b tell you about the relation?

Checkpoint

The tables that follow show variables changing in a pattern of exponential growth.

I.

x	0	1	2	3	4	5	6
y	1	2	4	8	16	32	64

II.

x	0	1	2	3	4	5	6
y	3	6	12	24	48	96	192

ⓐ What equation relating *NOW* and *NEXT* shows the common pattern of growth in the tables?

ⓑ How are the patterns of change in the tables different? How will that difference show up in plots of the tables?

ⓒ What equations ($y = \ldots$) will give rules for the patterns in the tables?

ⓓ How do the numbers used in writing those rules relate to the patterns of entries in the table? How could someone who knows about exponential growth examine the equation and predict the pattern in a table of (x, y) data?

Be prepared to share your equations and observations with the entire class.

On Your Own

Jurassic Park is a book and a movie about a dinosaur theme park. It is based on the idea that dinosaur DNA might be recovered from fossils and copied in laboratories until the genetic material for dinosaurs is available. The possibility of actually "recreating" dinosaurs is remote. But chemists *have* invented processes for copying genetic materials. The 1993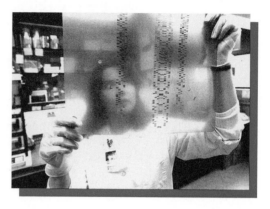

Nobel Prize for chemistry was shared by two scientists who developed such processes. In the PCR (polymer chain reaction) process invented by Kary Mullis, a sample of DNA is doubled. The process takes about 5 minutes.

a. Suppose a chemist starts the PCR process with a sample that holds only 7 copies of a special piece of DNA.

■ Write two different equations that can be used for calculating the number of copies of the DNA on hand after any number of 5-minute periods.

■ Use your equations from above to find the number of copies of the DNA produced after 2 hours.

■ Use your equations to find the number of 5-minute periods required to first produce 1 billion copies of the DNA.

b. How would your answers to Part a change if the starting DNA sample held 1, 2, or 3 copies of the DNA to be copied?

c. From your earlier study of linear models, you know that the exponential growth pattern common in living organisms is not the way all things change. For example, think about a car that accelerates quickly to the speed limit of 55 mph on a highway and keeps going at that speed for some time.

■ How long will it take the car to cover a distance of 250 miles?

■ What equations allow you to calculate the distance traveled by this car for any time? How are those equations different from what you expect to find with exponential growth?

■ What patterns do you expect to find in tables and graphs of (*time, distance*) data for this car? How are those patterns different from what you find with exponential growth?

MORE

Modeling

1. Suppose a single bacterium lands in an open cut on your leg and begins doubling every 15 minutes.

   a. How many bacteria will there be after 15, 30, 45, 60, and 75 minutes have elapsed (if no bacteria die)?

   b. Write rules that can be used to calculate the number of bacteria in the cut after any number of 15-minute periods.

      ■ Make the first an equation relating *NOW* and *NEXT*.

      ■ Make the second a rule using exponents, beginning "$y = \ldots$."

   c. Use your rules from Part b to make a table showing the number of bacteria in the cut at the end of each 15-minute period over 3 hours. Then describe the pattern of change in number of bacteria from each quarter hour to the next.

   d. Use the rules from Part b to find the predicted number of bacteria after 5, 6, and 7 hours. (**Hint:** How many 15-minute periods will that be?)

2. Suppose the wealthy Persian king in Investigation 2 offered his soldier an even more generous deal. The king will distribute 5 grains of rice for the first square, 25 for the second, 125 for the third, 625 for the fourth, and so on.

   a. Use an equation relating *NOW* and *NEXT* rice grain counts to find the number of grains of rice for each of the first 5 squares of the chessboard.

   b. Write a rule using exponents that could be used to calculate the number of grains of rice for any square, without starting from the first square.

   c. Use the rule in Part b to make a table showing the number of grains of rice for each of the first 10 squares. Describe the pattern of change in this table, from one square to the next.

   d. How would your answers to Parts a–c change if the king offered 5 grains for square 1, 10 grains for square 2, 15 grains for square 3, 20 grains for square 4, and so on?

3. The following sketches show several stages in the growth of a *fractal tree*.

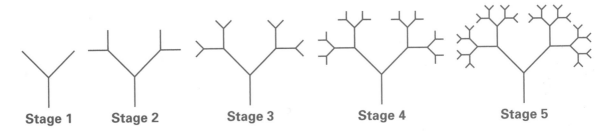

Stage 1   Stage 2   Stage 3   Stage 4   Stage 5

a. Suppose at Stage 0 there is one branch. Each year the tree grows exactly two new branches at the end of each branch. Make a table showing the number of new branches in each year from 0 to 10.

b. How many new branches will there be on this tree in year 20?

c. At what age will this sort of tree first produce at least 1 billion new branches?

d. Explain how you could find the answers to Parts a–c in three different ways using your calculator or computer.

e. Describe any symmetries that the fractal tree has at each stage. How could this information be used in drawing the next stage of the tree?

4. The drug penicillin was discovered by observation of mold growing on biology laboratory dishes. Suppose a mold begins growing on a lab dish. When first observed, the mold covers only $\frac{1}{8}$ of the dish surface, but it appears to double in size every day. When will the mold cover the entire dish?

Alexander Fleming, discoverer of penicillin

Organizing

1. Write each of the following calculations in shorter form using exponents.

a. $5 \times 5 \times 5 \times 5$

b. $3 \times 3 \times 3 \times 3 \times 3 \times 3 \times 3 \times 3$

c. $1.5 \times 1.5 \times 1.5 \times 1.5 \times 1.5 \times 1.5$

d. $(-10) \times (-10) \times (-10) \times (-10) \times (-10) \times (-10) \times (-10) \times (-10)$

e. $\underbrace{6 \times 6 \times \ldots \times 6}_{n \text{ factors}}$

f. $\underbrace{a \times a \times \ldots \times a}_{n \text{ factors}}$

2. Think about the meaning of an exponent as you complete the following tasks.

a. Do each of the following calculations without use of the exponent key ($\boxed{\wedge}$ or $\boxed{y^x}$) on your calculator.

i. 5^4 **ii.** $(-7)^2$ **iii.** 10^0

iv. $(-8)^3$ **v.** 2^8 **vi.** 2^{10}

b. Suppose b is any number and x is some positive whole number. Describe two ways in which you can calculate the value of b^x.

3. Exponential growth models, like linear models, can be expressed by an equation relating x and y values and an equation relating *NOW* and *NEXT* y values.

   a. Compare the patterns of (x, y) values produced by these two rules: $y = 2(3^x)$ and $y = 2 + 3x$.

   ■ For each rule, make a table of (x, y) values for x from 0 to 10 in steps of 1.

   ■ For each rule, plot the data obtained. The program TBLPLOT would be helpful here.

   ■ For each rule, write an equation using *NOW* and *NEXT* that could be used to produce the same pattern of (x, y) data.

   ■ For each rule, describe the way that y changes as x increases. Explain how that pattern shows up in the table and the graph.

   b. Now think about any two relations with rules $y = a(b^x)$ and $y = a + bx$ where $b > 1$.

   ■ What patterns are you sure to find in any table of (x, y) values in each case? What will the values of a and b tell about those patterns?

   ■ What patterns are you sure to find in graphs of the two relations? What will the values of a and b tell about those patterns?

   ■ What equations relating *NOW* and *NEXT* will give the same patterns of (x, y) values as the equations $y = a(b^x)$ and $y = a + bx$?

4. Shown below are partially completed tables for four relations between variables. In each case, decide if the table shows an exponential or a linear pattern of change. Based on that decision, complete the table as the pattern suggests. Then write equations for the patterns in two ways: using rules relating *NOW* and *NEXT* y values and using rules beginning "$y = \dots$" for any given x value.

a.

x	0	1	2	3	4	5	6	7	8
y				8	16	32			

b.

x	0	1	2	3	4	5	6	7	8
y				40	80	160			

c.

x	0	1	2	3	4	5	6	7	8
y				48	56	64			

d.

x	0	1	2	3	4	5	6	7	8
y				125	625	3,125			

5. For each pair of equations relating *NOW* and *NEXT y* values, produce tables and scatterplots of data. Then compare the patterns of growth by describing similarities and differences in the tables and graphs produced and in the rates of change.

   a. Compare change patterns produced by the equations *NEXT = NOW* × 3 and *NEXT = NOW* × 5, starting at 10 in each case.

   b. Compare change patterns produced by *NEXT = NOW* × 3 (starting at 5) and *NEXT = NOW* × 5 (starting at 3).

   c. Compare change patterns produced by *NEXT = NOW* × 3 (starting at 5) and *NEXT = NOW* + 3 (starting at 5).

   d. Compare change patterns produced by *NEXT = NOW* × 3 (starting at 5) and *NEXT = NOW* + 10 (starting at 100).

Reflecting

1. One common illness in young people is *strep throat*. This bacterial infection can cause painful sore throats. Have you or anyone you know ever had strep throat? How does what you have learned about exponential growth explain the way strep throat seems to develop very quickly?

2. Suppose you are asked to design a telephone calling tree for a school chorus that has 30 members. The purpose of the tree will be to help the director reach families of all chorus members as quickly and reliably as possible with information about trips, performances, and practices.

   a. Sketch diagrams of several different possible calling trees.

   b. Explain the advantages and disadvantages of each design.

3. You've now worked on many different problems involving exponential growth patterns.

   a. What are the key features of a relation between variables that are hints that exponential growth will be involved?

   b. How are the patterns of exponential growth models different from those of linear models?

4. Which of the two models for growth by doubling do you prefer: *NEXT = NOW* × 2 or $y = 2^x$? Give reasons for your preference and explain how the two models are related to each other.

5. The population of our world is now about six billion. At the present rate of growth, that population will double approximately every 50 years.

   a. If this rate continues, what will the population be 50, 100, 150, and 200 years from now?

   b. How would that growth pattern compare to a pattern that simply added six billion people every 50 years?

   c. Do you think the population is likely to continue growing in the "doubling every 50 years" pattern? Explain your reasoning.

   d. What do you think the effect of rapid population growth will be on your life in the 21st century?

Extending

1. Here are five stages in growth of another fractal design called the *dragon fractal*.

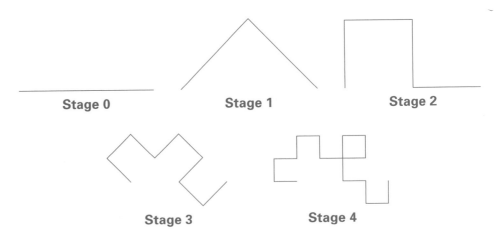

Stage 0   Stage 1   Stage 2

Stage 3   Stage 4

   a. Draw the next stage of growth in the dragon fractal.

   b. What pattern of change do you see in the number of segments of the growing fractal?

   c. Make a table and a plot of the data showing that pattern of change.

   d. Write an equation relating *NOW* and *NEXT* and an equation beginning "$y = \ldots$" for finding the number of segments in the figure at each Stage of growth.

   e. How many segments will there be in the fractal design at Stage 16?

   f. At what stage will the fractal design have more than 1,000 segments?

2. One of the most famous fractal forms is the *Koch snowflake*. It grows in much the same way as the tent-like fractal you explored in the "On Your Own" on page 427, except it starts with an equilateral triangle. In the first growing step, divide each segment into three equal pieces. Raise a "tent" over the center section with segments equal in length to the two remaining pieces on each side of the center section. Then the next stage repeats that process on each segment of the pattern at stage one, and so on.

a. Draw Stages 0–2 of the Koch snowflake.

b. Make a table showing the number of segments, the length of each segment, and the perimeter of the total figure at each stage through stage 6. Assume the length at Stage 0 is 1.

c. Write equations in two ways for each of the following relations: (*stage, number of segments*); (*stage, segment length*); (*stage, perimeter*).

d. Study the pattern in each variable (*number of segments, segment length,* and *perimeter*) as the number of growth stages increases to a very large number. Write a report describing your observations, making sure to comment especially on any surprising patterns.

3. Create a fanciful fractal of your own. Use color, or simply create it in black and white. Draw at least five stages of your fractal. Analyze it mathematically. Be sure to include a table, a graph, an equation relating *NOW* and *NEXT*, and an equation beginning "*y* = … ."

4. In this task, you will examine more closely the tent-like fractal from page 427.

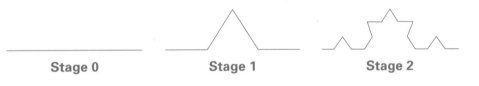

Stage 0   Stage 1   Stage 2

Recall that in moving from one stage to the next, each segment is divided into three equal-length parts. A tent is raised over the middle section with sides equal in length to the parts on each side.

a. If the original line segment is one inch long, how long is each segment of the pattern in Stage 1?

b. How long is each segment of the pattern in Stage 2?

c. Complete the following table showing the length of segments in the first ten stages.

Stage	0	1	2	3	4	5	6	7	8	9
Length	1	$\frac{1}{3}$	$\frac{1}{9}$							

d. Look back to Parts c and d of the "On Your Own" on page 427 where you wrote equations giving the number of short segments at each stage of the pattern. Then use that information and the results of Part c to complete the following table giving the length of the total pattern at each stage.

Stage	0	1	2	3	4	5	6	7	8	9
Length	1	$\frac{4}{3}$	$\frac{16}{9}$							

e. What appears to be happening to the length of the total pattern as the number of segments in the pattern increases?

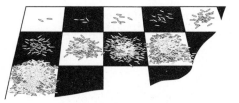

5. In the king's chessboard problem described in Investigation 2, it was easy to calculate the number of grains of rice for any given square. In this task, you will investigate the total number of grains for all squares taken together.

a. Find these sums:

$1 + 2 + 4 + 8$

$1 + 2 + 4 + 8 + 16$

$1 + 2 + 4 + 8 + 16 + 32$

$1 + 2 + 4 + 8 + 16 + 32 + 64$

b. What pattern do you see in the results of Part a that would allow you to predict the sum of any number of terms of this sequence? Test your conjecture on the sum: $1 + 2 + 4 + 8 + 16 + 32 + 64 + 128 + 256 + 512 + 1,024$. Revise your conjecture and test again if necessary.

c. Plan and carry out an investigation that would allow you to quickly calculate the sum of terms in a tripling sequence: $1 + 3 + 9 + 27 + 81 + 243 + ... + N$.

d. Try to find a pattern in your work in Parts b and c that would allow you to quickly calculate the sum of terms in any exponential sequence: $1 + r + r^2 + r^3 + r^4 + ... + r^n$.

Exponential Decay

In 1989, the oil tanker Exxon Valdez ran aground in waters near the Kenai peninsula of Alaska. Over 10 million gallons of oil spread on the waters and shoreline of the area, endangering wildlife. That oil spill was eventually cleaned up—some of the oil evaporated, some was picked up by specially equipped boats, and some sank to the ocean floor as sludge.

For scientists planning environmental cleanups, it is important to be able to predict the pattern of dispersion in such contaminating spills. *Think about* the following experiment that simulates pollution of a lake or river by some poison and the cleanup.

- Mix 20 black checkers (the pollution) with 80 red checkers (the clean water).

- On the first "day" after the spill, remove 20 checkers from the mixture (without looking at the colors) and replace them with 20 red checkers (clean water). Count the number of black checkers remaining. Then shake the new mixture. This simulates a river draining off some of the polluted water and a spring or rain adding clean water to a lake.

- On the second "day" after the spill, remove 20 checkers from the new mixture (without looking at the colors) and replace them with 20 red checkers (more clean water). Count the number of black checkers remaining. Then stir the new mixture.

- Repeat the remove/replace/mix process for several more "days."

The graphs below show two possible outcomes of the pollution and cleanup simulation.

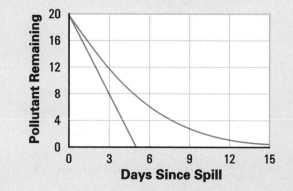

a What pattern of change is shown by each graph?

b Which graph shows the pattern of change that you would expect for this situation? Test your idea by running the experiment several times and plotting the (*time, pollutant remaining*) data.

c What sort of equation relating pollution *P* and time *t* would you expect to match your plot of data? Test your idea using a graphing calculator or computer.

The pollution cleanup experiment gives data in a pattern that occurs in many familiar and important problem situations. That pattern is called **exponential decay**.

INVESTIGATION 1 More Bounce to the Ounce

Most popular American sports involve balls of some sort. In playing with those balls, one of the most important factors is the bounciness or *elasticity* of the ball. For example, if a new golf ball is dropped onto a hard surface, it should rebound to about $\frac{2}{3}$ of its drop height.

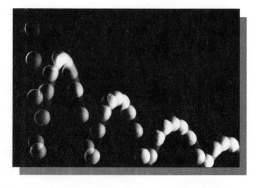

Suppose a new golf ball drops downward from a height of 27 feet onto a paved parking lot and keeps bouncing up and down, again and again.

1. Make a table and plot of the data showing expected heights of the first ten bounces.

Bounce Number	0	1	2	3	4	5	6	7	8	9	10
Rebound Height	27										

   a. How does the rebound height change from one bounce to the next? How is that pattern shown by the shape of the data plot?

   b. What equation relating *NOW* and *NEXT* shows how to calculate the rebound height for any bounce from the height of the preceding bounce?

   c. Write an equation beginning "$y = \ldots$" to model the rebound height after any number of bounces.

   d. How will the data table, plot, and equations for calculating rebound height change if the ball drops first from only 15 feet?

As is the case with all mathematical models, data from actual tests of golf-ball bouncing will not match exactly the predictions from equations of ideal bounces. You can simulate the kind of quality control testing that factories do by running some experiments in your classroom.

2. Get a golf ball and a tape measure or meter stick for your group. Decide on a method for measuring the height of successive rebounds after the ball is dropped from a height of at least 8 feet. Collect data on the rebound height for successive bounces of the ball.

   a. Compare the pattern of your data to that of the model that predicts rebounds which are $\frac{2}{3}$ of the drop height. Would a rebound height factor other than $\frac{2}{3}$ give a better model? Explain your reasoning.

   b. Write an equation using *NOW* and *NEXT* that relates the rebound height of any bounce of your tested ball to the height of the preceding bounce.

   c. Write an equation beginning "$y = \ldots$" to predict the rebound height after any number of bounces.

3. Repeat the experiment of Activity 2 with some other ball such as a tennis ball or a basketball.

   a. Study the data to find a reasonable estimate of the rebound height factor for your ball.

   b. Write an equation using *NOW* and *NEXT* and an equation beginning "$y = \ldots$" that model the rebound height of your ball on successive bounces.

Different groups might have used different balls and dropped the balls from different initial heights. However, the patterns of (*bounce number*, *rebound height*) data should have some similar features.

a Look back at the data from your two experiments.

- How do the rebound heights change from one bounce to the next in each case?
- How is the pattern of change in rebound height shown by the shape of the data plots in each case?

b List the equations relating *NOW* and *NEXT* and the rules ($y = \ldots$) you found for predicting the rebound heights of each ball on successive bounces.

- What do the equations relating *NOW* and *NEXT* bounce heights have in common in each case? How, if at all, are those equations different and what might be causing the differences?
- What do the rules beginning "$y = \ldots$" have in common in each case? How, if at all, are those equations different and what might be causing the differences?

c What do the tables, graphs, and equations in these examples have in common with those of the exponential growth examples in the beginning of this unit? How, if at all, are they different?

Be prepared to share and compare your data, models, and ideas with the rest of the class.

On Your Own

When dropped onto a hard surface, a brand new softball should rebound to about $\frac{2}{5}$ the height from which it is dropped. If a foul-tip drops straight down onto concrete after achieving a height of 25 feet, what pattern of rebound heights can be expected?

a. Make a table and plot of predicted rebound data for 5 bounces.

b. What equation relating *NOW* and *NEXT* and what rule ($y = \ldots$) giving height after any bounce match the pattern of rebound heights?

c. Here are some data from bounce tests of a softball dropped from a height of 10 feet.

Bounce Number	1	2	3	4	5
Rebound Height	3.8	1.5	0.6	0.2	0.05

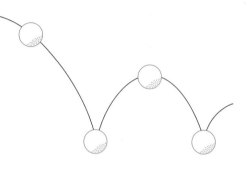

- What do these data tell you about the quality of the tested softball?

- What bounce heights would you expect from this ball if it were dropped from 20 feet instead of 10 feet?

d. What equation would model rebound height of an ideal soft-ball if the drop were from 20 feet?

INVESTIGATION 2 Sierpinski Carpets

One of the most interesting and famous fractal patterns is named after the Polish mathematician Waclaw Sierpinski. The first two stages in forming that fractal are shown here.

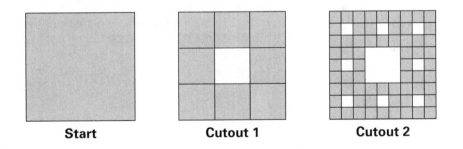

Start   **Cutout 1**   **Cutout 2**

Starting with a solid square "carpet" one meter on a side, smaller and smaller squares are cut out of the original in a sequence of steps. Notice how, in typical fractal style, small pieces of the design are similar to the design as a whole.

At the start of a Sierpinski carpet there is one square meter of carpet. But as cutting proceeds, there seems to be less and less carpet, and more and more hole.

1. Make a sketch showing the new holes that will appear in the third cutout from the carpet.

2. The carpet begins with an area of 1 square meter.

   a. How much of the original carpet is left after the first cutout?

   b. What fraction of the carpet left by the first cutout remains after the second cutout? How much of the original 1 square meter of carpet remains after the second cutout?

   c. What fraction of the carpet left by the second cutout remains after the third cutout? How much of the original 1 square meter of carpet remains after the third cutout?

   d. Following the pattern in the first three stages, how much of the original 1 square meter of carpet will remain after cutout 4? After cutout 5?

3. Write an equation showing the relation between the area of the remaining carpet at any stage and the next stage.

   a. What area would you predict for the carpet left after cutout 10?

   b. Find the area of the carpet left after cutout 20. After cutout 30.

4. Write an exponential equation that would allow you to calculate the area of the remaining carpet after any number of cutouts x, without going through all the cutouts from 1 to x.

   a. Make a table giving the area of the Sierpinski carpet from the start through cutout 10. Use TBLPLOT or similar software to make a plot of this data.

   b. How many cutouts are needed to get a Sierpinski carpet in which there is more hole than carpet remaining?

Checkpoint

Summarize the ways in which the table, graph, and equations for the Sierpinski carpet pattern are similar to, and different from, those for the following patterns:

ⓐ the bouncing ball patterns of Investigation 1;

ⓑ the calling tree, king's chessboard, and bacteria growth patterns of Lesson 1.

Be prepared to share your summaries of similarities and differences with the entire class.

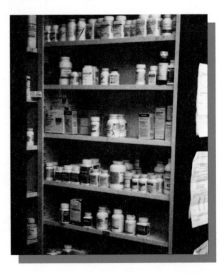

On Your Own

Suppose you started working on a very large Sierpinski carpet—a square that is 3 meters long on each side. Its starting area would be 9 square meters.

a. Find the area of the remaining carpet after each of the first 10 cutouts.

b. Make a plot of the (*cutout number, area*) data from Part a.

c. Write an equation that shows the change in area from one cutout to the next.

d. Write an exponential equation showing how to calculate the area of the carpet after any number *x* of cutouts.

e. How many cutouts are needed to get a Sierpinski carpet in which there is more hole than carpet remaining?

  ■ Show how the answer to this question can be found in a table of (*cutout number, area*) data.

  ■ Show how the answer to this question can be found in a plot of (*cutout number, area*) data.

f. How do your answers to Parts a–e compare to those for the first Sierpinski carpet with an original area of 1 square meter?

INVESTIGATION ▶ 3 Medicine and Mathematics

Drugs are a very important part of the human health equation. Many drugs are essential in preventing and curing serious physical and mental illnesses.

Diabetes, a disorder in which the body cannot metabolize glucose properly, affects people of all ages. In 1998, there were about 10 million diagnosed cases of diabetes in the United States. It was estimated that another 5 million cases remained undiagnosed.

In 5–10% of the diagnosed cases, the diabetic's body is unable to produce insulin, which is needed to process glucose.

To provide this essential hormone, these diabetics must take injections of a medicine containing insulin. The medications used (called insulin delivery systems) are designed to release insulin slowly. The insulin itself breaks down rather quickly. The rate varies greatly between individuals, but the following graph shows a typical pattern of insulin decrease.

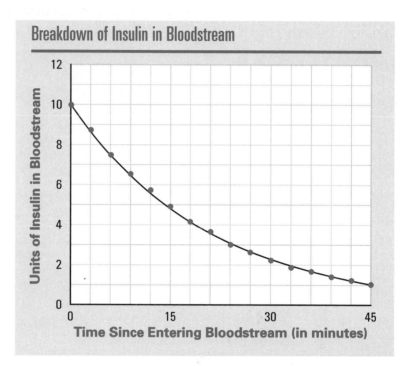

Breakdown of Insulin in Bloodstream

1. Medical scientists usually are interested in the time it takes for a drug to be reduced to one half of the original dose. They call this time the **half-life** of the drug. What appears to be the half-life of insulin in this case?

2. The pattern of decay shown on this graph for insulin can be modeled well by the equation $y = 10(0.95)^x$. Experiment with your calculator or computer to see how well a table of values and graph from this rule fit the pattern in the given graph. Then explain what the values 10 and 0.95 tell about the amount of insulin in the bloodstream.

3. What equation relating *NOW* and *NEXT* shows how the amount of insulin in the blood changes from one minute to the next, once 10 units have entered the bloodstream?

4. The insulin graph shows data points for each minute following the original insulin level. But the curve connecting those points reminds us that the insulin breakdown does not occur in sudden bursts at the end of each minute! It occurs *continuously* as time passes.

What would each of the following calculations tell about the insulin decay situation? Based on the graph on the previous page, what would you expect as reasonable values for those calculations?

a. $10(0.95)^{1.5}$ **b.** $10(0.95)^{4.5}$ **c.** $10(0.95)^{18.75}$

5. Mathematicians have figured out ways to do calculations with fractional or decimal exponents so that the results fit in the pattern for whole number exponents. One of those methods is built into your graphing calculator or computer software.

a. Enter the rule Y = 10(0.95^X) in the "Y=" list of your calculator or computer software. Then complete the following table of values showing the insulin decay pattern at times other than whole minute intervals.

Time in Minutes	0	1.5	4.5	7.5	10.5	13.5	16.5	19.5
Units of Insulin in Blood	10							

b. Compare the entries in this table with data shown by points on the graph on page 446.

c. Use your rule to estimate the half-life of insulin.

Checkpoint

In this unit, you have seen that patterns of exponential change can be modeled by equations of the form $y = a(b^x)$.

ⓐ What equation relates *NOW* and *NEXT* y values of this model?

ⓑ What does the value of *a* tell about the situation being modeled? About the tables and graphs of (x, y) values?

ⓒ What does the value of *b* tell about the situation being modeled? About the tables and graphs of (x, y) values?

ⓓ How is the information provided by values of *a* and *b* in exponential equations like $y = a(b^x)$ similar to, and different from, that provided by *a* and *b* in linear equations like $y = a + bx$?

Be prepared to compare your responses with those from other groups.

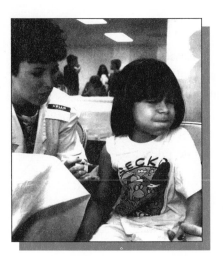

The most famous antibiotic drug is penicillin. After its discovery in 1929, it became known as the first *miracle drug*, because it was so effective in fighting serious bacterial infections.

Drugs act somewhat differently on each person, but on average, a dose of penicillin will be broken down in the blood so that one hour after injection only 60% will remain active. Suppose a patient is given an injection of 300 milligrams of penicillin at noon.

a. Make a table showing the amount of that penicillin that will remain at hour intervals from noon until 5 PM.

b. Make a plot of the data from Part a. Explain what the pattern of that plot shows about the rate at which penicillin decays in the blood.

c. Write an equation of the form $y = a(b^x)$ that can be used to calculate the amount of penicillin remaining after any number of hours x.

d. Use the equation from Part c to produce a table showing the amount of penicillin that will remain at *quarter-hour* intervals from noon to 5 PM. What can you say about the half-life of penicillin?

e. Use the equation from Part c to graph the amount of penicillin in the blood from 0 to 10 hours. Find the time when less than 10 mg remain.

MORE

Modeling • Organizing • Reflecting • Extending

Modeling

1. If a basketball is properly inflated, it should rebound to about $\frac{1}{2}$ the height from which it is dropped.

a. Make a table and plot showing the pattern to be expected in the first ten bounces after a ball is dropped from a height of 10 feet.

b. At which bounce will the ball first rebound less than 1 foot? Show how the answer to this question can be found in the table and on the graph.

c. Write two different forms of equations that can be used to calculate the rebound height after many bounces.

d. How will the data table, plot, and equations change for predicting rebound height if the ball is dropped from a height of 20 feet?

2. The sketch below shows the start and two cutout stages in making a triangular Sierpinski carpet. Assume that the area of the original triangle is 3 square meters.

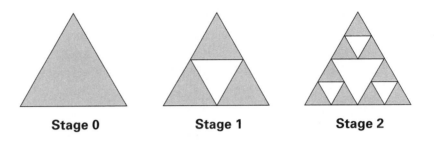

Stage 0   Stage 1   Stage 2

    a. Sketch the next stage in the pattern.

    b. Make a table showing (*cutout number, area remaining*) data for cutout stages 0 to 5 of this process.

    c. Make a plot of the data in Part b.

    d. Write two different equations that can be used to calculate the area of the remaining carpet at different stages. One equation should show change from one stage to the next. The other should be in the form "$y = \dots$."

    e. How many stages are required to reach the points where there is:

        ■ more hole than carpet remaining?

        ■ less than 0.1 square centimeter of carpet remaining?

    f. How are the pattern, table, graph, and equations for this triangular carpet similar to, and different from, those of the square carpets in Investigation 2?

    g. Describe any symmetries that the triangular carpet has at each stage.

3. You may have heard of athletes being disqualified from competitions because they have used anabolic steroid drugs to increase their weight and strength. These steroids can have very damaging side effects for the user. The danger is compounded by the fact that these drugs leave the human body slowly. With an injection of the steroid *ciprionate*, about 90% of the drug and its by-products will remain in the body one day later. Then 90% of that amount will remain after a second day, and so on. Suppose that an athlete tries steroids and injects a dose of 100 milligrams of ciprionate. Analyze the pattern of that drug in the athlete's body by completing the following tasks.

    a. Make a table showing the amount of the drug remaining at various times.

Time Since Use (in days)	0	1	2	3	4	5	6	7
Steroid Present (in mg)	100	90	81					

b. Make a plot of the data in Part a and write a short description of the pattern shown in the table and the plot.

c. Write two equations that describe the pattern of amount of steroid in the blood.

- Write one equation showing how the amount of steroid present changes from one day to the next.

- Write a second equation in the form $y = a(b^x)$ that shows how one could calculate the amount of steroid present after any number of days.

d. Use one of the rules in Part c to estimate the amount of steroid left after 0.5 and 8.5 days.

e. Estimate, to the nearest tenth of a day, the half-life of ciprionate.

f. How long will it take the steroid to be reduced to only 1% of its original level in the blood? That is, how many days will it take for only 1 milligram of the original dose to be left in the bloodstream?

4. When people suffer head injuries in accidents, emergency medical personnel sometimes administer a paralytic drug to keep the patient immobile. If the patient is found to need surgery, it's important that the immobilizing drug decay quickly.

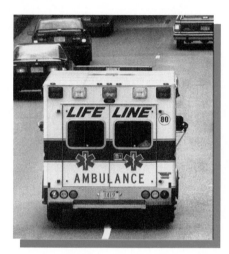

For one typical paralytic drug the standard dose is 50 micrograms. One hour after the injection, half the original dose has decayed into other chemicals. The halving process continues the next hour, and so on.

a. How much of the 50 micrograms will remain in the patient's system after 1 hour? After 2 hours? After 3 hours?

b. Write an equation for calculating the amount of drug that will remain x hours after the initial dose.

c. Use the equation from Part b to make a table showing the amount of drug left at half-hour intervals from 0 to 5 hours.

d. Make a plot of the data from Part c and then a continuous graph using the Y= and GRAPH commands.

e. How long will it take the 50-microgram dose to decay to less than 0.05 microgram?

5. In Unit 1, "Patterns in Data," you studied growth charts as you learned about percentiles. For children who fall under the 5th percentile level on these charts, a growth hormone may be used to help them grow at a more normal rate. If 10 milligrams of one particular growth hormone is introduced to the bloodstream, as much as 70% will still be present the next day. After another day, 70% of that amount will remain, and so on.

a. Write two different equations that can be used to calculate the amount of a 10-milligram dose of growth hormone remaining after any number of days.

b. How long will it take for the original 10-milligram dose to be reduced to 0.1 milligram? Show how the answer to this question can be found in a table of (*time, drug amount*) data and in a graph of that data.

c. What is the half-life of this growth hormone?

d. Suppose half the amount (5 milligrams) of the drug is introduced to the bloodstream. Compare the half-life of this dosage with that in Part c.

6. Radioactive materials have many important uses in the modern world, from fuel for power plants to medical x-rays and cancer treatments. But the radioactivity that produces energy and tools for "seeing" inside our bodies has some dangerous effects too; for example, it can cause cancer in humans.

The radioactive chemical strontium-90 is produced in many nuclear reactions. Extreme care must be taken in transportation and disposal of this substance. It decays rather slowly—if any amount is stored at the beginning of a year, 98% of that amount will still be present at the end of that year.

a. If 100 grams (about 0.22 pound) of strontium-90 are released due to an accident, how much of that radioactive substance will still be around after 1 year? After 2 years? After 3 years?

b. Write two different equations that can be used to calculate the amount of strontium-90 remaining from an initial 100 grams at any year in the future.

c. Make a table and a scatterplot showing the amount of strontium-90 that will remain from an initial amount of 100 grams at the end of every 10-year period during a century.

Years Elapsed	0	10	20	30	40	50	. . .
Amount Left (in g)	100						

d. Use one of the equations in Part b to find the amount of strontium-90 left from an initial amount of 100 grams after 15.5 years.

Limerick Generating Station, Montgomery County, PA. Owned and operated by PECO Energy Company.

e. Use one of the equations from Part b to find the number of years that must pass until only 10 grams remain.

f. Estimate to the nearest tenth of a year, the half-life of strontium-90.

Organizing

1. The following graphs, tables, and equations model four exponential growth and decay situations. For each graph, there is a matching table and a matching equation. Use what you know about the patterns of exponential relations to pair each graph with its corresponding table and equation. In each case, explain the clues that can be used to match the items without any use of a graphing calculator or computer.

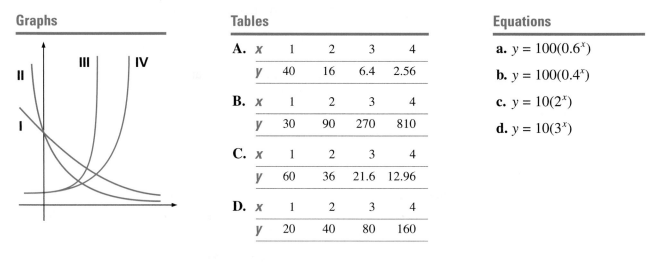

Graphs

Tables

A. x	1	2	3	4
y	40	16	6.4	2.56

B. x	1	2	3	4
y	30	90	270	810

C. x	1	2	3	4
y	60	36	21.6	12.96

D. x	1	2	3	4
y	20	40	80	160

Equations

a. $y = 100(0.6^x)$

b. $y = 100(0.4^x)$

c. $y = 10(2^x)$

d. $y = 10(3^x)$

2. This task will help you develop a better understanding of fractional or decimal exponents.

a. Use your calculator or computer software to produce a table of values for $y = 4^x$. Use values of x from 0 to 3 in steps of 0.1.

b. Use your calculator or computer software to produce a table of values for $y = (0.36)^x$. Use values of x from 0 to 4 in steps of 0.25.

c. Suppose you have an exponential model of the form $y = b^x$ where $b > 0$ and $b \neq 1$. If x is a decimal number between two numbers s and t, how is b^x related to b^s and b^t? Test your conjecture using exponential models different from those in Parts a and b.

d. Without using your calculator, estimate $3^{1.5}$. Explain the method you used.

3. For an equation of the form $y = a(b^x)$, what conclusions can you draw about the tables and the graphs of (x, y) values when b is

   a. between 0 and 1?

   b. greater than 1?

4. Suppose two equations $y = 100(b^x)$ and $y = 100(c^x)$ model the decay of 100 grams of two different radioactive substances. How can you tell which substance will have the shorter half-life by comparing the values of b and c? How will this difference appear in the graphs of the two equations?

Reflecting

1. Which example of an exponential decay pattern seems to you to be the most interesting or important example of exponential decay—the Sierpinski carpets, metabolizing of drugs in the body, bouncing of a golf ball, or decay of radioactive chemicals? Give reasons for your choice.

2. Suppose the makeup of a drug is such that one hour after a dose is administered to an individual's bloodstream, 70% remains active. Does it follow that one-half hour after administration of the same dose to the same person, 85% will remain active? Explain your reasoning.

3. When an exponential equation models a situation such as in the metabolizing of drugs in the body, fractional or decimal exponents are useful. In other situations, they may not be particularly meaningful.

   a. Give an example of an exponential growth or decay situation where use of fractional exponents would not make sense.

   b. What are some characteristics of exponential growth or decay situations that would suggest that use of fractional exponents is sensible?

4. Suppose a problem situation is modeled by the equation $y = 500(0.6^x)$. Tell as much as you can about the nature of the situation.

5. Suppose a person taking steroid drugs is hospitalized due to a side effect from the drug. Tests taken upon admittance show a steroid concentration of 1.0. The next test one day later shows a concentration of 0.75. Based on these results the person's family and friends assume that in three days the drug will be out of the person's system.

   a. What pattern of change are the family and friends assuming?

   b. What might be a more accurate pattern prediction? Why is this pattern more reasonable?

Extending

1. Fleas are one of the most common pests for dogs. If your dog has fleas, you can buy many different kinds of treatments, but those wear off over time. Suppose the half-life of one such treatment is 10 days.

   a. Make a table showing the fraction of an initial treatment that will be active after 0, 10, 20, 30, and 40 days.

   b. Experiment with your calculator or computer to find an equation of the form $y = b^x$ (where x is time in days) that matches the pattern in your table of Part a.

2. In Unit 3, "Linear Models," you solved linear equations of the form $ax + b = c$ and $ax + b = cx + d$ using several different methods. Solve each of the following *exponential equations* in at least two different ways.

   a. $2^x = 6$

   b. $2x = 2^x$

3. The following program creates a drawing of the Sierpinski Triangle (Modeling Task 2) in an interesting way. This program is from the *TI-83 Graphing Calculator Guidebook*.

Program: SIERPINS

```
:FnOff              :For(K,1,3000)      :.5(1+Y)→Y
:ClrDraw            :rand→N             :End
:PlotsOff           :If N≤1/3           :If 2/3<N
:AxesOff            :Then               :Then
:0→Xmin             :.5X→X              :.5(1+X)→X
:1→Xmax             :.5Y→Y              :.5Y→Y
:0→Ymin             :End                :End
:1→Ymax             :If 1/3<N and N≤2/3  :Pt - On(X,Y)
:rand→X             :Then               :End
:rand→Y             :.5(.5+X)→ X        :StorePic 6
```

Reprinted by Permission of Texas Instruments.

   a. Enter the program in your calculator.

   b. After you execute the program, recall and display the picture by pressing RecallPic 6.

   c. How is the idea of *NOW* and *NEXT* used in the design of this program?

Compound Growth

Every now and then we hear about somebody winning a big payoff in a state lottery somewhere. The winnings can be 1, 2, 5, or even as large as 50 million dollars. Those big money wins are usually paid off in annual installments for about 20 years. But some smaller prizes like $10,000 are paid at once. How would you react if this news report were actually about you?

Kalamazoo Teen Wins Big Lottery Prize—$20,000

A Kalamazoo teenager has just won $20,000 from a Michigan lottery ticket that she got as a birthday gift from her uncle. In a new lottery payoff scheme, the teen (whose name has been withheld) has two payoff choices: One option is to receive $1,000 payments each year for the next 20 years. In the other plan, the lottery will invest $10,000 in a special savings account that will earn 8% interest compounded annually for 10 years. At the end of that time she can withdraw the balance of the account.

Think About This Situation

Suppose you had just won the lottery.

a Which of the two payoff methods would you choose?

b Which method do you think would give the greatest total payoff?

c About how much money do you think would be in the special savings account at the end of 10 years?

INVESTIGATION 1 Just Like Money in the Bank

Of the two lottery payoff methods, one has quite a simple rule: $1,000 per year for 20 years, giving a total payoff of $20,000. The plan to put $10,000 in a savings account paying 8% **compound interest** might not be as familiar.

■ After one year your balance will be:
$$10,000 + (0.08 \times 10,000) = 1.08 \times 10,000 = 10,800.$$

■ After the second year your balance will be:
$$10,800 + (0.08 \times 10,800) = 1.08 \times 10,800 = 11,664.$$

During the next year the savings account balance will increase in the same way, starting from $11,664, and so on.

1. Write equations that will allow you to calculate the balance of this deposit

   a. for any year, using the balance from the year before.

   b. after any number of years *x*.

2. Use the equations to make a table and a plot showing the growth of this special savings account for a period of 10 years.

Time (in years)	0	1	2	3	4	...	9	10
Balance ($)	10,000	10,800	11,664					

3. Describe the pattern of growth in this savings account as time passes.

   a. Why is the balance not increasing at a constant rate?

   b. How could the pattern of increase be predicted from the shape of the graph of the modeling rules?

4. How long would it take to double the $10,000 savings account?

5. Compare the pattern of change and the final account balance in Activity 2 to that for each of the following possible savings plans over 10 years. Write a summary of your findings.

   a. Initial investment of $15,000 earning only 4% annual interest.

   b. Initial investment of $5,000 earning 12% annual interest.

Checkpoint

Most savings plans operate in a manner similar to the special lottery savings account. They may have different starting balances, different interest rates, or different periods of investment.

a Describe two ways to find the value of such a savings account at the end of each year from the start to year 10. Use methods based on

   ■ an equation relating *NOW* and *NEXT*.

   ■ an exponential equation $y = a(b^x)$.

b What is the shape of the graphs that you would expect?

c How will the rules change as the interest rate changes? As the amount of initial investment changes?

d Why does the dollar increase in the account get larger from one year to the next?

Be prepared to explain your methods and ideas to the entire class.

On Your Own

The world population and populations of individual countries grow in much the same pattern as money earning interest in a bank.

    i. Brazil is the most populous country in South America. In 2000, its population was about 170 million. It is growing at a rate of about 1.5% per year.

    ii. Nigeria is the most populous country in Africa. Its 2000 population was about 123 million. It is growing at a rate of about 2.8% per year.

a. Assuming these growth rates continue, make a table showing the predicted populations of these two countries in each of the 10 years after 2000. Then make a scatterplot of the data for each country.

    ■ Describe the patterns of growth expected in each country.

    ■ Explain how the different patterns of growth are shown in the scatterplots.

b. Write equations to predict the populations of these countries for any number of years x in the future. Use the equations

    ■ to estimate when Brazil's population might reach 300 million.

    ■ to estimate when Nigeria's population might reach 200 million.

MORE

Modeling • Organizing • Reflecting • Extending

Modeling

1. Suppose that a local benefactor wants to offer college scholarships to every child born into a community. When a child is born, the benefactor puts $5,000 in a special savings fund earning 5% interest per year.

    a. Make a table and graph of an account showing values each year for 18 years.

    b. Compare the pattern of growth of the account in Part a to one in which the initial deposit was $10,000. Compare values of each account after 18 years.

c. Compare the pattern of growth of the account in Part a to one in which the interest rate was 10% and the initial deposit was $5,000. Compare values of each account after 18 years.

d. Compare values of the accounts in Parts b and c after 18 years. Explain why your finding makes sense.

2. In 2000, the population of Iraq was 23.1 million and was growing at a rate of about 2.8% per year, one of the fastest growth rates in the world.

a. Make a table showing the projected population of Iraq in each of the eight years after 2000.

b. Write two different kinds of equations that could be used to calculate population estimates for Iraq at any time in the future.

c. Estimate the population of Iraq in 2020.

d. What factors might make the estimate of Part c an inaccurate forecast?

3. In Unit 2, "Patterns of Change," you studied growth in the population of Arctic bowhead whales. The natural growth rate was about 3.1% and estimates place the 1993 population between 6,900 and 9,200. The harvest by Inuit people is very small in relation to the total population. Disregard the harvest for this task.

a. If growth continued at 3.1%, what populations would be predicted for each year to 2010? Make tables based on both 1993 population estimates.

b. How would the pattern of results in Part a change if the growth rate were and continued to be 7%, as some scientists believe it is?

c. Write two different types of equations that can be used to calculate population estimates for the different possible combinations of initial population and growth rate estimates.

d. Find the likely time for the whale population to double in size under each set of assumptions.

Organizing

1. Consider the four exponential equations:

   i. $y = 5(1.2^x)$ ii. $y = 5(1.75^x)$

   iii. $y = 5(2.5^x)$ iv. $y = 5(3.25^x)$

   a. Sketch the patterns of graphs you expect from these four equations; then check your estimates with your calculator or computer.

b. Make tables of (x, y) values for the four equations and explain how the patterns in those tables fit the shape of the graphs in Part a.

2. Think about the shapes of graphs of exponential equations.

   a. Sketch the graph shape you would expect from an exponential equation $y = a(b^x)$ when $0 < b < 1$. Sketch the shape you would expect when $b > 1$.

   b. How does the value of a affect the graph?

3. One way to think about rates of growth is to calculate the time it will take for a quantity to double in value. For example, it is common to ask how long it will take a bank investment or a country's population to double.

   a. If the U.S. population in 2000 was about 276 million and growing exponentially at a rate of 0.9% per year, how long will it take for the U.S. population to double?

   b. One year's growth is 0.9% of 276 million, or about 2.5 million. How long would it take the U.S. population to double if it increased *linearly* at the rate of 2.5 million per year?

   c. How long does it take a bank deposit of $5,000 to double if it earns interest compounded annually at the rate of 2%? At a 4% rate? At a 6% rate? At an 8% rate? At a 12% rate?

   d. Examine your (*rate, time to double*) data in Part c. Do you see a pattern that would allow you to predict the doubling time for an investment of $5,000 at an interest rate of 3% compounded annually? Check your prediction. If your prediction was not close, search for another pattern for predicting doubling time and check it.

4. What property of addition and multiplication justifies each of the following calculations in figuring compound interest?

   - $10,000 + 0.06(10,000) = 1.06(10,000)$
   - $10,600 + 0.06(10,600) = 1.06(10,600)$
   - $P + rP = (1 + r)P$

Reflecting

1. What characteristic of money earning interest in the bank and growth of human or animal populations makes them grow in similar exponential patterns?

2. Which of these two offers would you take to invest a $500 savings? Justify your choice.

   i. 4% interest paid each year on the balance in that year

   ii. $20 interest paid each year (Notice: $20 = 4% of $500.)

3. Refer to the savings fund description in Modeling Task 1 and complete Part a if you have not already done so. Calculate *differences* between the value of the account at the beginning and end of years 1, 5, 11, and 18. What do these four differences say about the rate at which the savings account grows?

Extending

1. Banks frequently pay interest more often than once each year. Suppose your bank pays interest compounded *quarterly*. If the annual percentage rate is 4%, then the bank pays 1% interest at the end of each 3-month period.

   a. Explore the growth of a $1,000 deposit in such a bank over 5 years.

   b. Compare the quarterly compounding with annual compounding at 4%.

   c. Repeat the calculations and comparisons if the annual rate is 8%.

2. Many people borrow money from a bank to buy a car, a home, or to pay for a college education. However, they have to pay back the amount borrowed plus interest. To consider a simple case, suppose that for a car loan of $9,000 a bank charges 6% annual rate of interest compounded quarterly and the repayment is done in quarterly installments. One way to figure the balance on this loan at any time is to use the equation:

   new balance = 1.015 × *old balance* – *payment*

   a. Use this equation to find the balance due on this loan for each quarterly period from 0 to 20, assuming that the quarterly payments are all $250.

   b. Experiment with different payment amounts to see what quarterly payment will repay the entire loan in 20 payments (5 years).

   c. Experiment with different loan amounts, interest rates, and quarterly payments to see how those factors are related to each other. Write a brief report of your findings.

3. The value of purchased products such as automobiles *depreciates* from year to year.

   a. Would you suspect the pattern of change in value of an automobile from year to year is linear? Exponential?

   b. Select a 1995 automobile of your choice. Research the initial cost of the car and its value over the years since 1995.

   c. Does the plot of (*time since purchase*, *value*) data show an exponential pattern? If so, find an exponential model that fits the data well. Use your model to predict the value of the car when it is 10 years old.

   d. Compare your findings with those of other classmates who completed this task. Which 1995 automobile researched held its value best? Explain your reasoning.

Lesson 4 ▶ Modeling Exponential Patterns in Data

Our planet Earth is home for millions of different species of animal and plant life. Many species of life become *extinct* every day. Much loss is due to human actions that change the environment for animal and plant life. The endangerment of life forms often can be reversed by protection of species and their habitats.

For example, the Kenai Peninsula of Alaska is a natural home for wolves, moose, bear, and caribou. When a gold rush in 1895–96 brought thousands of prospectors to the area, hunting and changes to the environment reduced all those populations. Wolves disappeared from the Kenai Peninsula by about 1915.

In the late 1950s a few wolves reappeared on the Kenai Peninsula. They were protected from hunting in 1961. The graph at the right shows the growth of the Kenai wolf population during the decade after it became a protected species.

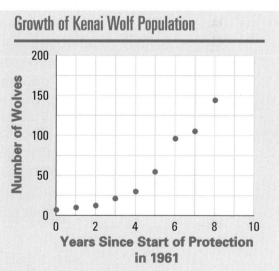

Growth of Kenai Wolf Population

Think About This Situation

Study the pattern of change in the Kenai wolf population.

a What sort of graph model do you think would best fit the trend in wolf population data?

b How would you find equations modeling the growth of the wolf population on the Kenai Peninsula over time?

c How might a Natural Resources Department use your equations?

INVESTIGATION ▶ 1 Popcorn and Thumbtacks

This investigation includes four experiments that give you practice in modeling exponential patterns in data. In each case, you will collect some data showing how a variable changes over time. Then you will make a table and a scatterplot of that data. Next you will experiment to find an equation relating *x* and *y* to fit the pattern in your data. Finally, you will use the equation to make predictions. The main steps are outlined in Experiment 1. Apply the same steps in the other three experiments.

Share the work so that each group does Experiment 1 and one of the remaining three experiments. Compare results to see similarities and differences in outcomes of the same and different experiments.

Experiment 1

Begin this experiment with a paper plate divided into four equal sections. Shade one of the sections, as shown here. You also need a paper cup containing 200 kernels of unpopped popcorn.

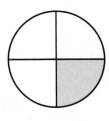

Pour the kernels of corn from the cup onto the plate at its center. Shake the plate gently to spread the kernels around. Then *remove* and count all kernels that end up on the shaded sector. (This ends Stage 1 of the experiment.) For the next stage, repeat the shake-remove process with the kernels that remain.

1. **Collect Data** Collect the data from your experiment in a table like this:

Stage Number	0	1	2	3	4	5	6	7
Kernels Left	200							

2. **Display Data** Make a scatterplot of the (*stage*, *kernels left*) data.

3. **Analyze Data** Write a short description of the pattern in the data table and scatterplot.

   a. Experiment with entering different equations in the "Y=" function list. See if you can find one that fits the pattern well.

   b. Use the **exponential regression** feature of your calculator or computer software to find the equation of the best-fitting exponential model.

c. Write the exponential equation $y = a(b^x)$ that you believe fits your data best.

d. Explain how the values of a and b relate to the experiment you have done. What do they tell about stages and kernels of corn?

4. Apply Model Use your exponential equation to do the following:

a. Predict the number of kernels of corn left after Stage 3 and Stage 7. Compare your predictions with the data table. Explain any differences you observe.

b. Write a question about the experiment that can be answered by using your modeling equation. Then show how you would use the equation to find the answer.

Experiment 2

Begin this experiment by pouring 20 kernels of corn onto the plate that has been divided into quarters. After shaking the plate to spread the kernels, count the number of kernels that land on the shaded sector of the plate. *Add that number of kernels to the test supply.* (This ends Stage 1.) Repeat the shake-count-add process several times.

Record the data in a table of (*stage, supply of kernels after adding*) data. Plot that data. Find an equation that models the relation between stage number and number of kernels. Explain the relation between the equation and the experiment. Compare values predicted by your model with data table entries. Write a question about the experiment. Use your equation to answer the question.

Experiment 3

Begin this experiment with a collection of 200 identical thumbtacks. Toss those tacks onto a flat surface and *remove* all tacks that land with the point up. Count and record the number of tacks remaining. (This ends Stage 1.)

Point Up   Point Down

Toss the remaining tacks again and *remove* all tacks that land with their point up. Record the number of tacks left. Repeat the toss-remove-count process several times.

Record the data in a table of (*stage, tacks left*) data. Plot that data. Find an equation that models the relation between stage number and number of tacks. Explain the relation between the equation and the experiment. Compare values predicted by your model with data table entries. Write a question about the experiment. Use your equation to answer the question.

Experiment 4

Begin this experiment by tossing 10 tacks onto a flat surface. Count the number of tacks that land with their point down and *add* that number of tacks to the test supply. (This ends Stage 1.) Repeat the toss-count-add process several times.

Record the data in a table of (*stage, supply of tacks after adding*) data. Plot that data. Find an equation that models the relation between the stage and the number of tacks. Explain the relation between the equation and the experiment. Compare values predicted by your equation with data table entries. Write a question about the experiment. Use your equation to answer the question.

Checkpoint

After each experiment has been reported to your class, consider the following questions about exponential patterns in data, their graphs, and their equations $y = a(b^x)$.

a What does the value of a indicate about each of the experiments? What does it indicate about the table of values for any exponential relation?

b What does the value of b indicate about each of the experiments? What does it indicate about the table of values for any exponential relation?

c In what sorts of problem situations can you expect an exponential model for which the value of b is between 0 and 1?

d In what sorts of problem situations can you expect an exponential model for which the value of b is greater than 1?

e What changes in data, graphs, and equation models would you expect if the experiments were changed

- by dividing the paper plate differently in the corn kernel experiment?
- by using different tacks, with shorter or longer points, in the tack experiment?

Be prepared to explain your ideas to the entire class.

On Your Own

Suppose you have a collection of 300 dice. You toss them and remove all that show ones or sixes. Then you count and record the number of dice remaining. You repeat this procedure until you have only a few dice left.

a. What pattern of (*toss number, dice remaining*) data would you expect?

b. What would a scatterplot of the (*toss number, dice remaining*) data look like?

c. What equation would you expect as a good model of the data?

d. Suppose your teacher gives a homework assignment to conduct the dice-tossing experiment for ten tosses. If you were the teacher, what would you think:

■ If two students reported exactly the same data?

■ If one student's data fit exactly the equation $y = a(b^x)$ that was reported as the most likely model for that data pattern?

INVESTIGATION ▶ 2 Another Day Older ... and Deeper in Debt

By the end of 1995, the national debt of the United States government was about $5,000,000,000,000. The debt was growing by nearly $900,000,000 every day. The numbers may seem too large to comprehend and the national debt may seem unrelated to your life. But on February 17, 1993, the *Chicago Tribune* reported that a large group of young people had gathered to make their position on the national debt known. These young people were concerned about government spending in this country and its long range implications. How will the national debt affect your financial future? Is the damage to the financial health of future generations beyond repair?

Danziger © 1993 *The Christian Science Monitor*

1. Consider the data table below which gives the federal debt in trillions of dollars from 1988 to 1993.

Year	1988	1989	1990	1991	1992	1993
Debt in $ Trillion	2.6	2.9	3.2	3.6	4.0	4.4

Source: U.S. Bureau of the Census, *Statistical Abstract of the United States: 1995* (115th edition). Washington, DC, 1995.

a. Enter these data in your calculator or computer using 0 for 1988, 1 for 1989, and so on. Make a scatterplot.

b. What patterns in the data and the plot suggest that a linear model might fit the trend in national debt? What patterns in the data are unlike linear models?

c. Use your calculator or computer software to find a linear rule for the (*year*, *debt*) data that you believe fits well. Find an exponential rule that seems to fit well, also.

- Use the two rules to make tables and graphs. Describe the similarities and differences.

- Use each model equation to estimate the national debt in the years 2005, 2010, 2020, and 2030.

d. Which of the two models do you think is better for describing and predicting the American national debt? Compare your choice with those of other groups. Resolve any differences between groups.

2. Early estimates of the national debt data for 1994 indicated that the rate of increase might be slowing. The debt figure for 1994 actually turned out to be around 4.6 trillion. Subsequent data, shown in the table below, confirm that the rate of increase is slowing.

Year	1995	1996	1997	1998	1999	2000
Debt in $ Trillion	4.97	5.22	5.41	5.5	5.66	5.67

Source: U.S. Bureau of the Census, *Statistical Abstract of the United States: 2000.* Washington, DC, 2000.

a. How do these additional values affect the pattern of the data plot?

b. How might they affect equations of good linear and exponential models?

c. How might they affect the long-range projection of the national debt?

d. Do you think today's young people have as much need to be concerned about the national debt as those who gathered in 1993?

Consider your studies of tables and graphs of data from two related variables.

ⓐ What patterns suggest use of a linear model? What patterns suggest an exponential model?

ⓑ If you think a linear model is probably a good one to use, how can you find values of a and b for the modeling equation $y = a + bx$?

ⓒ If you think an exponential model is probably a good one to use, how can you find values of a and b for the modeling equation $y = a(b^x)$?

Be prepared to share your thinking and methods for fitting models to data with the entire class.

On Your Own

Earthquakes are among the most damaging kinds of natural disasters. The size of an earthquake is generally reported as a rating on the *Richter scale*—usually a number between 1 and 9. That Richter scale rating indicates the energy released by the shaking of the ground and the height of the shock waves recorded on seismographs.

The data in the following table show Richter scale ratings and amounts of energy released for six earthquakes.

Earthquake Location	Richter Scale Rating	Energy (in sextillion ergs)
San Francisco, CA, 1906	8.25	1,500
Yugoslavia, 1963	6.0	0.63
Alaska, 1964	8.6	5,000
Peru, 1970	7.8	320
Italy, 1976	6.5	3.5
Loma Prieta, CA, 1989	7.1	28

a. Use your calculator or computer software to make a scatterplot for these data.

b. What pattern makes it reasonable to think that this is an exponential relation?

c. Use your calculator or computer software to find an algebraic model that fits the data pattern well.

d. Use the model to estimate energy released by earthquakes listed in the following chart.

Earthquake Location	Richter Scale Rating	Energy (in sextillion ergs)
Quetta, India, 1906	7.5	
Kwanto, Japan, 1923	8.2	
Chillan, Chile, 1939	7.75	
Agadir, Morocco, 1960	5.9	
Iran, 1968	7.4	
Tangshan, China, 1976	7.6	
Northridge, CA, 1994	6.7	
Kobe, Japan, 1995	7.2	

MORE
Modeling • Organizing • Reflecting • Extending

Modeling

1. Suppose you are given a collection of 200 new pennies and directed to perform this experiment: Shake the pennies and drop them on a flat surface. *Remove* all pennies that land heads up. Count and record the number of pennies remaining. Repeat the shake-drop-remove-count process several times.

  a. What pattern of (*drop number, pennies left*) data would you expect from this experiment?

  b. What pattern would you expect in a scatterplot of the data?

  c. What equations would you expect to be good models of the data?

  d. Conduct the experiment to test your predictions.

2. Suppose you are given a collection of new pennies and directed to perform this experiment: Start with a cup holding 10 pennies. Shake the pennies and drop them on a flat surface. Count the pennies that turn up tails. *Add* that number of pennies to your cup and record the number of pennies now in the cup. Repeat the shake-drop-count-add process several times.

  a. What pattern of (*drop number, number of pennies*) data would you expect from this experiment?

b. What pattern would you expect in a scatterplot of the data?

c. What equations would you expect to be good models of the data?

d. Conduct the experiment to test your predictions.

3. Try this experiment with a supply of about 200 plastic spoons. Toss the spoons onto a flat surface and *remove* all spoons that land right side up. Count and record the number of spoons remaining. Repeat this toss-remove-count process several times.

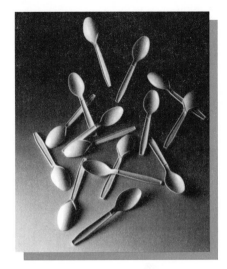

a. Record the (*toss number, number of spoons*) data in a table.

b. Make a scatterplot of the data.

c. Find an equation that models the relation between toss number and number of spoons. Explain the relation between the equation and the experiment.

d. Write a question about this experiment. Use your model to answer the question.

4. Try this spoons experiment. Start with a supply of about 15 plastic spoons. Toss the spoons onto a flat surface and count the spoons that land right side up. *Add* that number of new spoons to your test collection. Then repeat the toss-count-add process several times.

a. Record the (*toss number, number of spoons*) data in a table.

b. Plot the data.

c. Find an equation that models the relation between toss number and number of spoons.

d. Explain the relation between the equation and the experiment. How is the equation related to the experiment in Modeling Task 3?

5. With improved health care and advances in medicine, people continue to live longer. The American Hospital Association has predicted that the nursing home population will increase rapidly. It has made the following projections.

Projected Nursing Home Population

Year	1985	1990	2000	2010	2020	2030	2040	2050
Population (millions)	1.30	1.45	1.80	2.30	2.55	3.35	4.30	4.80

Source: Person, J.E. Jr., ed., *Statistical Forecasts of the United States*. Detroit: Gale Research, Inc. 1993.

a. Find linear and exponential models that fit these data well. Use 0 for 1985, 5 for 1990, 10 for 1995, and so on.

b. In the equation $y = a + bx$ for the linear model, what do the values of a and b indicate about the projected pattern of change in nursing home populations?

c. In the equation $y = a(b^x)$ for the exponential model, what do the values of a and b indicate about the projected pattern of change in nursing home populations?

d. Which model do you believe was used to make the projections in the table?

e. What is the projected number of elderly who will be receiving nursing home care when you are eighty years old?

6. Life began on earth millions of years ago. Our species, *Homo Sapiens*, dates back only 300,000 years. The black rhinoceros, the second largest of all land mammals, has walked the earth for 40,000,000 years. In less than a century, the very existence of this species has been threatened. Prior to the 19th century, over 1,000,000 black rhinos roamed the plains of Africa. That number has been drastically reduced by hunting over the years. Recent data on the black rhino population is shown in the table below.

African Black Rhino Population					
Year	1970	1980	1984	1986	1993
Population (in 1000s)	65	15	9	3.8	2.3

Source: Nowak, R.M., *Mammals of the World*, fifth ed., vol. 2. Johns Hopkins University Press: Baltimore, 1991; www.rhinos-irf.org/rhinos/black.html

a. Make a scatterplot of these data and find an exponential equation that models the pattern in the data well. Use 0 for 1970, 10 for 1980, and so on.

b. Use the model from Part a to predict the black rhino population in the year 2010.

c. Use your model to find the year when the black rhino population is predicted to be less than 1,000.

d. Find a linear model for the black rhino data that you believe fits the data well. Answer Parts b and c again using that model.

e. A model based on very few data points is sometimes inaccurate, especially if one data point has an incorrect value. Suppose the 1970 black rhino population was actually only 30,000. Find what you believe is a good-fitting model in that case.

f. The actual black rhino population in 2000 was approximately 2,700. What might account for the break in the pattern suggested by earlier data?

Organizing

1. Without using your calculator or computer, sketch graphs for each of the following equations. Explain the reasoning you used in making each sketch.

   a. $y = 3x + 5$

   b. $y = 5(3^x)$

   c. $y = 5\left(\frac{1}{3}\right)^x$

2. Make tables of sample (x, y) data that fit the conditions below. Use values for x from 0 to 8. Explain your reasoning in making each table.

   a. The y values increase exponentially from an initial value of 5.

   b. The y values increase exponentially from an initial value of 5 at a greater rate than the example in Part a.

   c. The y values increase linearly from an initial value of 5.

   d. The y values decrease exponentially from an initial value of 25.

   e. The y values decrease linearly from an initial value of 25.

3. Complete a table like the one below so that it shows:

   a. a pattern of linear growth. Write a linear equation that describes the pattern.

   b. a pattern of exponential growth. Write an exponential equation that describes the pattern.

x	0	1	2	3	4	5
y		10	20			

4. When a fair coin is flipped, the outcome of "heads" or "tails" is equally likely. So the probability of a flipped coin landing heads up is $\frac{1}{2}$ or 0.5. Refer back to Experiment 3 of this lesson.

   a. If one of the thumbtacks is tossed in the air, what do you think is the probability that it will land with the point up?

   b. Toss a thumbtack in the air 100 times and count the number of times the tack lands with the point up. Use the results of your experiment to get a better estimate of the probability in Part a.

   c. What do the results of your experiment in Part b tell you about the probability that a tack tossed in the air will land point down?

   d. How, if at all, is the probability of a tack landing point up reflected in the equation model for Experiment 3?

Reflecting

1. A thousand-dollar bill is about 0.0043 inches thick. Imagine a stack of thousand-dollar bills whose total value is a trillion dollars. How high would the stack of bills be? If you created a stack of thousand-dollar bills whose value is that of the current national debt, about how many miles high would the stack be? Does this help show the seriousness of the debt you and your classmates will be inheriting?

2. Jail overcrowding is an issue in many states. Drug use and drug-related crime have contributed to the problem. Average operating costs of $25,000 per inmate and construction costs of $50,000 per cell will be an incredible burden on these jail systems. Examine the data in the following table, which gives the total number of jail inmates for the years 1993 through 1999.

U. S. Jail Inmates (excluding federal and state prisons)							
Year	1993	1994	1995	1996	1997	1998	1999
Jail Population	459,804	486,474	507,044	518,492	567,079	592,462	605,943

Source: *Statistical Abstract of the United States: 2000.* Washington, DC: U.S. Bureau of the Census, 2000.

   a. What sort of model seems best for projecting this growth pattern into the future—linear, exponential, or some other type?

b. Assume that the jails in these states were all full in 1999. Make a reasonable estimate of the number of additional cells that will be needed by 2005. Estimate the construction and operating costs for the additional cells.

c. Where might the money for the increased costs come from?

3. Health care spending has been another factor in American life which has shown exponential growth. Using the data below, create a scatterplot and find an algebraic model to closely fit this data.

Total U.S. Spending on Health Care

Year	1975	1980	1985	1990	1995	1998
Spending in $ Billion	132	247	429	699	993	1,149

Source: *Statistical Abstract of the United States: 2000.* Washington, DC: U.S. Bureau of the Census, 2000.

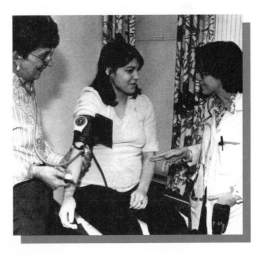

a. Predict the health care spending total for 2005 and 2010.

b. Besides inflation, what factors do you think would cause this dramatic rise in the cost for health care?

c. "Cost per capita" indicates the expense per person and therefore adjusts to reflect changes in population. The population of the U.S. for the given years is indicated in the following table. Calculate the health care cost per capita.

U.S. Population

Year	1975	1980	1985	1990	1995	1998
Population in Millions	216	228	238	250	263	271

Source: *Statistical Abstract of the United States: 2000.* Washington, DC: U.S. Bureau of the Census, 2000.

d. Is the cost per capita for health care also growing exponentially? What implications would this have for the future of health care in this country?

Extending

1. The following data were collected using a Geiger counter and a sample of radioactive barium-137. A Geiger counter measures the level of radioactivity in a sample.

Geiger Counter Readings for Barium-137

Time in Minutes	0	1	2	3	4	5	6
Counts per Minute	10,094	8,105	5,832	4,553	3,339	2,648	2,035

a. Make a scatterplot of the data.

b. Find an exponential model that fits the data well and use it to estimate the half-life of this radioactive substance—the time when only half of an original amount is predicted to remain.

2. The following data is from the *HIV/AIDS Surveillance Report*.

AIDS Cases and Fatalities

Year	1995	1996	1997	1998	1999
Estimated Population Living with AIDS	216,796	240,184	268,242	293,702	320,282
Fatalities Adults/Adolescents	49,284	50,070	37,356	21,704	17,806
Fatalities Children < 13 years	542	431	219	124	113

Source: U.S. Department of Health and Human Services. *HIV/AIDS Surveillance Report 12, No. 2* (July 2000).

a. Make a scatterplot for each of the three sets of data included in the table, showing how those variables have changed over time. Describe the patterns in the scatterplots.

b. Find a model you believe will help you make the best predictions about the number of persons living with AIDS in the future. Use your model to predict the number of persons who will be living with AIDS in 2010.

c. Find a model for the child AIDS fatalities data. Assuming the pattern will continue, use your model to predict the number of AIDS fatalities for children in the year 2010.

d. Suppose you were making the same prediction in 1986 using the earlier data given in the table below:

Year	1981	1982	1983	1984	1985
AIDS Fatalities Children < 13 years	9	13	29	48	115

Source: U.S. Department of Heath and Human Services. *HIV/AIDS Surveillance Report 5, no. 2* (July 1993).

- ■ What model would you have used based on only the data available at that time?

- ■ Does that model give you the same prediction for AIDS fatalities for children in 2010 as your model from Part c?

- ■ What changes in conditions might explain the differences?

3. Cigarette smoke contains nicotine, a very addictive and harmful chemical that affects the brain, nervous system, and lungs. The productivity losses and health care costs associated with cigarette smoke are considerable.

Suppose an individual smokes one cigarette every 40 minutes over a period of three hours and that each cigarette introduces 100 units of nicotine into the bloodstream. The half-life of nicotine is 20 minutes.

a. Create a chart that keeps track of the amount of nicotine which remains in the body over the three-hour time period in 20-minute intervals. Plot these totals over time. Then describe the pattern of nicotine build-up in the body of a smoker.

b. How would the data change if the individual smokes a cigarette every 20 minutes?

c. Because nicotine is a very addictive drug, it is difficult for a smoker to break the habit. Suppose a long-time smoker decides to quit "cold turkey." That is, rather than reducing the number of cigarettes smoked each day, the smoker resolves never to pick up another cigarette. How will the level of nicotine in that smoker's bloodstream change over time?

4. Alcohol is another dangerous drug. Driving after excessive drinking is not only punishable by law but also potentially fatal. The National Highway Safety Administration reported 2,238 youth alcohol-related traffic fatalities in 1999— an average of 6 fatalities each day.

While legal limits of blood alcohol concentration (BAC) are different in each state, the American Medical Association recommends that a limit of 0.05 be used.

There are many factors that affect a person's BAC. Some factors include body weight, gender, and the amount of alcohol drunk. The following chart contains typical data relating body weights and number of drinks consumed to approximate blood alcohol concentration. (Because of individual differences, this chart should not be considered to apply to everyone.)

Approximate Blood Alcohol Concentration

Weight (in pounds)	1 Drink	2 Drinks	3 Drinks	4 Drinks	5 Drinks
100	0.05	0.09	0.14	0.18	0.23
120	0.04	0.08	0.11	0.15	0.19
140	0.03	0.07	0.10	0.13	0.16
160	0.03	0.06	0.09	0.11	0.14
180	0.03	0.05	0.08	0.10	0.13

Source: National Highway Traffic Safety Administration, *Driving under the influence: A report to Congress on alcohol limits*. Washington, DC., 1992.

a. As time passes since alcohol was consumed, a person's body metabolizes the drug. Again, the rate at which this happens is different for each person. For most people, their BAC would drop at a rate of at least 0.01 each hour. Suppose a 120 lb. person had consumed three drinks. Using the table above and a burnoff rate of 0.01 per hour, when would this person satisfy the AMA's suggested limit of 0.05? How would this change if the person weighed 180 pounds and had consumed four drinks?

b. Prepare a graphical display of the data in the chart. Describe any patterns you see in the display.

c. Create a table showing the change in blood alcohol level over time, for a 140 lb. person who has consumed five drinks. Make a scatterplot of this data.

d. What type of model would best fit the scatterplot in Part c?

e. Based on your work in Investigation 3 of Lesson 2 and in Extending Task 3 on the previous page, what type of model best describes the amount of substance in the body over time for steroids, nicotine, and penicillin?

Looking Back

Many interesting and important patterns of change involve quantities changing as time passes. Populations of animals, bacteria, and plants grow over time. Drugs in the blood and radioactive chemicals in the environment decay over time. Most people hope that their bank savings account grows quickly over time. In many of the examples, the change is modeled well by exponential rules of the form $y = a(b^x)$.

When you find an exponential model for change in a variable, that model can be used to make useful predictions of events in the future. Test your understanding of exponential models on the following problems.

1. Code numbers are used in hundreds of ways every day—from student and social security numbers to product codes in stores and membership numbers in clubs.

   a. How many different 2-digit codes can be created using the digits 0, 1, 2, 3, 4, 5, 6, 7, 8, and 9 (for example, 33, 54, 72 or 02)?

   b. How many different 3-digit codes can be created using those digits?

   c. How many different 4-digit codes can be created using those digits?

   d. Using any patterns you may see, complete a table like the one below showing the relation between number of digits and number of different possible codes.

Number of Digits	1	2	3	4	5	6	7	8	9
Number of Codes									

   e. Write an equation using *NOW* and *NEXT* to describe the pattern in the table of Part d.

   f. Write an equation that shows how to calculate the number of codes C for any number of digits D used.

g. Kitchenware stores stock thousands of different items. How many digits would you need in order to have code numbers for up to 8,500 different items?

h. How will your answers to Parts a through f change if the codes were to begin with a single letter of the alphabet (A, B, C, … , or Z) as in A23 or S75?

2. In one professional golf tournament, the money a player wins depends on her finishing place in the standings. The first place finisher wins $\frac{1}{2}$ of the $1,048,576 in total prize money. The second place finisher wins $\frac{1}{2}$ of what is left; then the third place finisher wins $\frac{1}{2}$ of what is left, and so on.

U.S. Women's Open Champion
Annika Sorenstam

a. What fraction of the *total* prize money is won

- by the second place finisher?

- by the third place finisher?

- by the fourth place finisher?

b. Write a rule showing how to calculate the share of the prize money won by the player finishing in *n*th place, for any *n*.

c. Make a table showing the actual prize money in dollars (not fraction of the total prize money) won by each of the first ten place finishers.

Place	1	2	3	4	5	6	7	8	9	10
Prize (dollars)										

d. Write a rule showing how to calculate the actual prize money in dollars won by the player finishing in place *n*. How much money would be won by the 15th place finisher?

e. How would your answers to Parts a through d change if

- the total prize money were reduced to $500,000?

- the fraction used were $\frac{1}{4}$ instead of $\frac{1}{2}$?

f. When prize monies are awarded using either fraction, $\frac{1}{2}$ or $\frac{1}{4}$, could the tournament organizers end up giving away more than the stated total prize amount? Explain your response.

3. Growth of protected wild animal populations like the Alaskan wolves can be simulated as follows:

- Assume that the population starts with 4 adult wolves, 2 male and 2 female.

- Assume that each year, each female produces 4 pups who survive (assume 2 male and 2 female survivors in each litter). Thus, at the end of the first year there will be 12 wolves (6 male and 6 female). At the end of the next year, there will be 36 wolves (18 male and 18 female), and so on.

a. In what ways does this seem a reasonable simulation of the population growth? What modeling assumptions seem unlikely to be accurate?

b. Make a table showing the number of wolves at each stage (assume no deaths).

Stage	0	1	2	3	4	5	6	7	8
Wolf Count	4	12	36						

c. Use your calculator or computer software to find both linear and exponential models for the data in your table. Compare the fit of the two models to the data pattern. Explain which you feel is the better model.

d. What patterns of change will occur in the graph and in the table of values of a linear model? Of an exponential model? How do those typical patterns help you to decide which model is best in Part c?

e. Why does the wolf population grow at a faster rate as time passes?

f. How would the numbers in your table change if you assumed that wolves lived only 5 years? How does that affect the growth rate of the population?

4. In a study of ways that young people handle money, four high school students were given $200 at the start of a school year. They were asked to keep records of what they did with that money for the next 10 months.

- Cheryl put the money away for safe keeping and worked so she could add 10% to the total every month.

- James put his money in a box at home and added $10 each month.

- Jennifer put her money in a box at home and spent $10 each month.

- Delano put his money in a box at home. At the start of each month, he took out 10% of his balance for spending in that month.

a. The following four graphs show possible patterns for the savings of the students over time. Match each student's saving or spending pattern to the graph that best fits it. Explain your reasoning.

i. **ii.**

iii. **iv.**

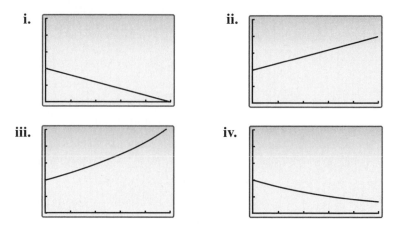

b. Match each graph from Part a to the type of rule you would expect to produce it, $y = a + bx$ or $y = a(b^x)$. Then explain what you can expect for values of a and b in each rule.

5. If x is a whole number, calculations like 2^x, 3^x or $\left(\frac{1}{2}\right)^x$ involve many multiplications. For example, $2^{10} = 2 \times 2 \times 2 \times 2 \times 2 \times 2 \times 2 \times 2 \times 2 \times 2$. On a calculator, you can reduce the number of operations by using an exponential key such as $\boxed{\wedge}$ or $\boxed{y^x}$. What could you do on a basic four-function calculator that has no exponential key? Answer Parts a through d assuming you have only number keys, an = ($\boxed{\text{ENTER}}$) key, and the multiplication key ($\boxed{\times}$).

a. How could you calculate 2^{10} with fewer than nine $\boxed{\times}$ keystrokes (no addition allowed)?

b. How could you use the fact that $2^{10} = 1{,}024$ to calculate 2^{20} with only one $\boxed{\times}$ keystroke?

c. How could you use the fact that $2^{10} = 1{,}024$ and $2^5 = 32$ to calculate 2^{15} with only one $\boxed{\times}$ keystroke?

d. How could you calculate 3^{12} in several different ways with only the $\boxed{\times}$ key?

e. How could you calculate $3^{12} \times 3^8$ with only an exponential key?

f. Look back at the results of your work on Parts a–e for a pattern that will complete calculations of this type: $2^m \times 2^n = 2^?$ for any nonnegative, whole-number values of n and m. Explain the rule your group invents using the meaning of exponents.

g. Explain why the rule you came up with in Part f also applies when the base 2 is replaced by 3 or 6 or 1.5 or any other positive number.

Exponential models can be used to solve problems in many different situations.

a In deciding whether an exponential model will be useful, what hints do you get from

■ the patterns in data tables?

■ the patterns in graphs or scatterplots?

■ the nature of the situation and the variables involved?

b Exponential models, like linear models, can be expressed by an equation relating x and y values and by an equation relating *NOW* and *NEXT* y values.

■ Write a general rule for an exponential model, $y = \ldots$.

■ Write a general equation relating *NOW* and *NEXT* for an exponential model.

■ What do the parts of the equations tell you about the situation being modeled?

c How can the rule for an exponential model be used to predict the pattern in a table or graph of that model?

d How are exponential models different from linear models?

e What real situations would you use to illustrate exponential change for someone who did not know what those patterns are like and used for?

Be prepared to share your ideas, equations, and examples with the whole class.

On Your Own

Write, in outline form, a summary of the most important mathematical concepts and methods developed in this unit. Organize your summary so that it can be used as a quick reference in future units and courses.

Simulation Models

Simulating Chance Situations

In some cultures, it is customary for a bride to live with her husband's family. Therefore, couples who have no sons and whose daughters all marry will have no one to care for them in their old age.

Customs of a culture and the size of its population often lead to issues that are hard to resolve. China had a population of over 1,200,000,000 in 2000. In an effort to reduce the growth of its population, the government of China instituted a policy to limit families to one child. The policy has been very unpopular among rural Chinese who depend on sons to carry on the family farming.

Think About This Situation

The situation described above raises many interesting mathematical questions as well as societal ones.

a In a country where parents are allowed to have only one child, what is the probability that their one child will be a son? What is the probability they will not have a son? What assumption(s) are you making when you answer these questions?

b If each pair of Chinese parents really had only one child, do you think the population would increase, decrease, or stay the same? Explain your reasoning.

c Describe several alternative plans that the government of China might use to control population growth. For each plan, discuss how you might find the answers to the following questions.

■ What is the probability that parents will have a son?

■ What will happen to the total population of China?

■ What will rural couples think about your plan?

INVESTIGATION 1 How Many Children?

In Part c of the "Think About This Situation" on the previous page, you shared different ways to examine the effects of your policy. In real life, it is hard to gather data that easily show the effects of a policy on the population. It may take several generations to see the long term effects. To estimate these effects you can *simulate* the situation in a way that allows informative data to be gathered more easily and quickly. In this investigation, you will simulate situations by flipping a coin.

1. Suppose China implements a new policy that allows each family to have two children.

   a. Explain how to use a coin to simulate the birth of *one* child. What did a head represent? What did a tail represent? What assumption(s) are you making?

   b. Explain how to use a coin to simulate the births of *two* children to a family. What are the possible outcomes?

   c. When you simulate a family with two children by flipping a coin twice and recording the results, you have conducted one **trial**. To be sure you have a reasonable estimate of what two-child families look like, it is necessary to conduct many trials. Conduct 200 trials simulating two-child families. Share the work among the groups in your class. Make a frequency table like the one below to record the results of your 200 trials.

Type of Family	Frequency
Two Girls	
Older Girl and Younger Boy	
Older Boy and Younger Girl	
Two Boys	
Total Number of Trials	200

d. Use your frequency table to estimate the probability that a family with two children will have *at least one* boy.

e. Estimate the probability that a family with two children will have at least one boy using a mathematical method other than simulation. Explain your other method.

f. Do the four types of families—two girls, older girl and younger boy, older boy and younger girl, two boys—appear to be **equally likely**? Describe the meaning of *equally likely* for a friend who is not in this class.

g. What is the total number of children in the 200 trials in Part c? What is the total number of girls? Of boys?

Here is one plan for reducing population growth that your class may have discussed.

Allow parents to continue to have children until a boy is born.
Then no more children are allowed.

For most of the remainder of this investigation, you will examine this plan. You will begin by making your best prediction about the effects of such a policy. Then you will use simulation techniques to improve your estimates.

2. Suppose that in rural China all parents continue having children until they get a boy. After the first boy, they have no other children. In your group, discuss each question below. Write your best prediction of the answer to each question.

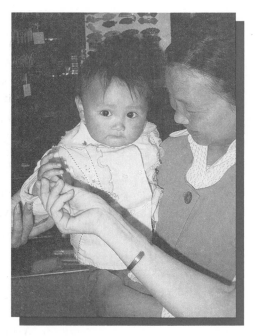

a. Will more boys or more girls be born in rural China?

b. What will be the average number of children per family in rural China?

c. Will the population of rural China increase, decrease, or stay the same?

d. What percentage of families will have only one child?

e. What percentage of families will have four children or more?

f. What percentage of the children in rural China will belong to single-child families?

To get a good estimate of the answers to the questions in Activity 2, your group could simulate the situation. To do this, design a **simulation model** that imitates the process of parents having children until they get a boy.

3. Explain how to use a coin to conduct one trial that models a family having children until they get a boy.

   a. Carry out one trial for your simulation of having children until a boy is born. Make a table like the following one. Make a tally mark (/) in the frequency column opposite the number of tosses it took to get a "boy."

Number of Tosses to Get a "Boy"	Frequency		Number of Tosses to Get a "Boy"	Frequency
1			7	
2			8	
3			9	
4			10	
5			⋮	
6			**Total Number of Trials**	50

   b. Continue carrying out trials of having children until a boy is born. Stop when you have a total of 50 "families." Divide the work among the members of your group. Record your results in the frequency table. Add as many additional rows to the table as you need.

   ■ How many of your 50 families had four children or more?

   ■ How many boys were born in your 50 families?

   ■ How many girls were born in your 50 families?

4. Now use your frequency table to estimate answers to the six questions posed in Activity 2.

   a. Compare your estimates with your original predictions. For which questions did your initial prediction vary the most from the simulation estimate? (If most of your original predictions were not accurate, you are in good company. Most people aren't very good at predicting the answers to probability problems.)

   b. Write several misconceptions that you or others in your group originally had about this situation.

5. Make a histogram of your group's frequency table on a graph like the one shown below or on your calculator or computer.

   a. Describe the shape of this distribution.

   b. What is the largest family size? The smallest?

   c. On the histogram, locate the median and the lower and upper quartiles of the distribution.

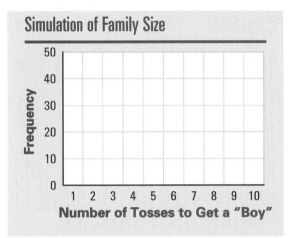

6. Each group should reproduce their histogram on a sheet of paper or on the chalkboard.

   a. Describe any similarities in the histograms.

   b. Describe any differences in the histograms. Explain why the differences occurred.

   c. Combine the frequency tables from all of the groups in your class into one frequency table on the overhead projector or chalkboard.

   d. Make a histogram of the combined frequency table. How are the histograms from the individual groups similar to this one? How are they different?

   e. Reproduced below are the questions from Activity 2. Estimate the answers to these questions using the combined frequency table of Part c above.

      ■ Will more boys or more girls be born in rural China?

      ■ What will be the average number of children per family in rural China?

      ■ Will the population of rural China increase, decrease, or stay the same?

      ■ What percentage of families will have only one child?

      ■ What percentage of families will have four children or more?

      ■ What percentage of the children in rural China will belong to single-child families?

   f. Should you have more confidence in the estimates from Part e or in the estimates from your group? Explain your reasoning.

7. In the "Think About This Situation" on page 484, your class proposed several alternative plans for reducing population growth in China.

   a. As a class, choose a plan different from the one in which parents have children until they get a boy, and design one trial of your plan.

   b. Perform at least 200 trials, sharing the work among groups in the class. Place your results in a frequency table.

   c. Under your plan, what is the probability that parents will have a son? How did you calculate the probability?

   d. Will the population of China increase, decrease, or remain the same under your plan? Explain your reasoning.

Checkpoint

In this lesson, you learned how to design simulations.

   a Describe, in your own words, what a *simulation model* is. Why is it important to conduct a large number of trials?

   b Why is it always a good idea to make a histogram of the results of a simulation?

   c Describe a way to simulate the have-children-until-you-get-a-boy plan that does not use coins.

   d If you flip a coin until you get a head and then repeat this many times, will you tend to have a larger proportion of heads or of tails?

   Be prepared to share your descriptions and thinking with the class.

Simulation is a good way to estimate the answer to a probability problem. The greater the number of trials, the more likely it is that the estimated probability is close to the actual probability. In our complex world, simulation is often the only feasible way to deal with a problem involving chance. Simulation is an indispensable tool to scientists, business analysts, engineers, and mathematicians.

"I've had it! Simulated wood, simulated leather, simulated coffee, and now simulated probabilities!"

STATISTICS: CONCEPTS AND CONTROVERSIES by Moore © 1979 by W.H. Freeman and Company. Used with permission.

▶ On Your Own

When asked in what way chance affected her life, a ninth-grader in a very large Los Angeles coeducational city high school noted that students are chosen randomly to be checked for weapons. Suppose that when this policy was announced, a reporter for the school newspaper suspected that the students would not be chosen randomly, but that boys would be more likely to be chosen than girls. The reporter then observed the first search and found that all 10 students searched were male.

a. If a student is in fact chosen randomly, what is the probability that the student will be a boy?

b. Write instructions for conducting one trial of a simulation that models selecting 10 students at random and observing whether each is a boy or a girl.

c. What assumptions did you make in your model?

d. Perform 20 trials using your simulation model.

e. Report the results in a frequency table showing the number of boys selected.

f. Write an article for a school paper describing your simulation, its results, and your conclusion. Include a histogram in your article.

Modeling • Organizing • Reflecting • Extending

Modeling

1. A new plan to control population growth in rural China is proposed. Parents will be allowed to have *at most three* children and must stop having children as soon as they get a boy.

   a. Describe how to conduct one trial that models one family that follows this plan.

   b. Conduct 5 trials. Copy the following frequency table which gives the results of 195 trials. Add your results to the frequency table so that there is a total of 200 trials.

Type of Family	Frequency
First Child is a Boy	97
Second Child is a Boy	50
Third Child is a Boy	26
Three Girls and No Boy	22
Total Number of Trials	

   c. Estimate the percentage of families that would not have a son.

   d. Make a histogram of the results in your frequency table.

   e. How does the shape of this histogram differ from that of the have-children-until-you-get-a-boy plan? Explain why you would expect this to be the case.

   f. What is the average number of children per family? Will the total population increase or decrease under this plan or will it stay the same?

   g. In the long run, will this population have more boys or more girls or will the numbers be about equal? Explain your reasoning.

2. Jeffrey is taking a 10-question true-false test. He didn't study and doesn't even have a reasonable guess on any of the questions. He answers "True" or "False" at random.

   a. With your group, decide how to conduct one trial that models the results of a true-false test.

   b. Conduct 50 trials. Share the work. Record your results in a frequency table that gives the number of questions Jeffrey got correct in each trial.

c. Make a histogram of the results in your frequency table. Describe its shape.

d. Use your frequency table to estimate, on average, the number of questions Jeffrey will get correct. Theoretically, what is the number of questions that he should expect to get correct using his random guessing method?

e. If 70% is required to pass the test, what is your estimate of the probability that he will pass the test?

3. The winner of the World Series of baseball is the first team to win four games.

a. What is the fewest number of games that can be played in the World Series? What is the greatest number of games that can be played in the World Series? Explain.

b. Suppose that the two teams in the World Series are evenly matched. Describe how to conduct one trial simulating a World Series.

c. Use your simulation model to determine the probability that the series lasts seven games. Conduct 5 trials. Construct a frequency table similar to the one shown below and add your 5 results so that there is a total of 100 trials.

Number of Games Needed in the Series	Frequency
4	11
5	21
6	30
7	33
Total Number of Trials	

d. Make a histogram of the results in your frequency table.

e. What is your estimate of the probability that the series will go seven games?

4. According to the U.S. Department of Education report, *The Condition of Education 2000* (page 42), about 50% of high school seniors say they like mathematics.

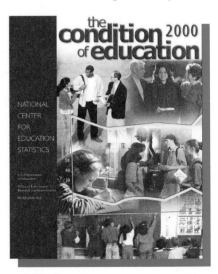

a. Describe how to conduct one trial of a simulation to answer the following questions.

- Would it be unusual to find that only 12 seniors of a randomly selected group of 30 said they liked mathematics?

- Would it be unusual for all 30 randomly selected seniors to say they like mathematics?

b. Conduct five trials using your simulation model. Copy the frequency table below that shows the results of 195 trials. Add your results to the table.

Number of Seniors Who Like Mathematics	Frequency
6	1
7	0
8	1
9	6
10	5
11	9
12	19
13	29
14	26

Number of Seniors Who Like Mathematics	Frequency
15	28
16	18
17	24
18	13
19	7
20	5
21	3
22	1

c. Use your results to estimate answers to the questions in Part a.

d. Select 30 seniors at your school and survey them to determine how many like mathematics. Is the number unusual? If so, what are possible reasons for this?

e. Suppose you survey a classroom of seniors selected at random in your school and find that all 30 of the students in the class like mathematics. List as many reasons as you can why this could occur.

Organizing

1. Refer to your class frequency table from Part c of Activity 6 (page 488).

   a. Make a scatterplot of the (*number of tosses to get a "boy"*, *frequency*) data.

   b. Would a line be a reasonable model of the relationship between number of tosses and frequency? Why or why not?

2. In 2000, China had a population of approximately 1,250,000,000. Assume parents were not following the one-child-per-family policy and the population of China continued to grow at about 1% per year.

   a. At that rate, how many people would have been added to the population of China in 2001? Compare this number to the population of your state.

   b. Assume the growth rate of 1% per year continues. Make a table of the year-end populations of China from 2000 to 2010.

   - Is the relationship a linear one? Explain your reasoning on the basis of your table.

   - What is the percent increase in China's population from 2000 to 2010?

   c. Make a scatterplot of the data in your table.

   - Find an equation that fits this data well.

   - Use your equation to predict when the population of China will exceed 1.5 billion people.

3. Suppose that during first period, Central High School has 95 classes of 30 students each and 5 classes of 100 students each.

   a. What is the average first period class size as reported by the high school?

   b. Suppose each student records the size of his or her first period class. What is the average class size from the students' point of view?

   c. How are these questions similar to Activity 2, Parts d and f of Investigation 1?

4. You can use random devices other than coins to simulate the birth of a child.

   a. How could a cube be used to simulate the birth of a child—either boy or girl?

   b. Could you use a regular tetrahedron to simulate the birth of a child? Explain your procedure.

   c. Identify other geometric shapes that could be used as the basis for a simulation model of births.

5. A circle with radius 6 inches is inscribed in a square as a model of a dartboard. Suppose a randomly thrown dart hits the board.

   a. What is a reasonable estimate of the probability the dart will land in the circle? Explain your reasoning.

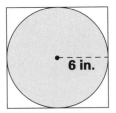

6 in.

b. Describe a simulation that could be used to estimate the probability.

c. Why might throwing darts at the board not be a good simulation?

d. How could you use the results of your simulation in Part b to estimate π?

Reflecting

1. Is there any difference in the following three questions? Explain your position.

   a. In a country where parents are allowed to have only one child, what percentage of couples will have a daughter?

   b. In a country where parents are allowed to have only one child, what is the probability that a couple, selected at random, will have a daughter?

   c. In a country where parents are allowed to have only one child, what fraction (proportion) of couples will have a daughter?

2. In future work in *Contemporary Mathematics in Context*, you will have to decide if trials are *independent*—past events don't change the probabilities.

   a. Suppose you toss a penny four times and get heads each time. What is the probability you will get a head on the fifth toss?

   b. If a family has four boys in a row, what is the probability the next child will be a girl? How is this question different from the one in Part a? How could you find the answer to this question?

3. Do some library research to find information about one of the following questions. Prepare a brief written report on your findings.

   a. In what cultures of the world is there a strong preference for male children? For female children? What effect does this have on children of the opposite gender in these cultures?

   b. Does a preference for male or female children exist in the United States? How do we know whether or not this is the case? If there is a preference for a particular gender in the United States, what might be some possible reasons for this preference?

4. A school is selling magazine subscriptions to raise money. A group wants to simulate the situation of asking ten people if they will buy a magazine. Jason proposes that the group flip a coin ten times and count the number of heads since a person either buys a magazine (heads) or doesn't buy a magazine (tails). Is Jason's simulation model a reasonable one? Explain your position.

5. In Part c of Activity 1 of Investigation 1 (p. 485), you simulated the situation of counting the number of girls and boys in a family that has two children. Does it matter if you flip *one* coin twice or flip *two* coins once? Explain your reasoning.

Extending

1. You have studied four plans in this lesson for population planning.

   ■ Each family has only one child.

   ■ Each family has exactly two children.

   ■ Each family has children until they have a boy (or girl).

   ■ Each family has at most three children, stopping when they get a boy (or girl).

   Suppose that all people marry at the age of 20 and have children immediately. Assume the first generation consists of an imaginary population of 32 twenty-year-old men and 32 twenty-year-old women. How many children can we expect to be in the fifth generation under each one of the four plans above? (Assume that in every generation there are equal numbers of males and females.)

2. If a student guesses on every question of a 10-question true-false test, how many questions would you expect that student to get correct? What percentage is this? Devise a method of scoring a true-false test so that a student who guesses on every question would expect to get a 0% on the test and a student who knows all of the answers gets 100%.

3. In 1974, 48 male bank supervisors were each given one personnel file and asked to judge whether the person should be promoted or the file held and other applicants interviewed. The files were identical except that half of them indicated that the file was that of a female and half indicated that the file was that of a male. Of the 24 "male" files, 21 were recommended for promotion. Of the 24 "female" files, 14 were recommended for promotion. (B. Rosen and T. Jerdee, "Influence of Sex Role Stereotypes on Personnel Decisions," *Journal of Applied Psychology*, Volume 59, 1974, pages 9–14.) Design and carry out a simulation to evaluate how likely it is that a difference this large could occur just by chance. After conducting your simulation, do you believe that the bank managers were discriminating or do you believe the different numbers of people promoted could reasonably have happened just by chance?

4. Look up the records of the number of games actually played in past World Series.

   a. Make a histogram of the number of games actually played per series.

   b. How does the shape of the histogram of the real data compare to the one of your simulated data in Part d of Modeling Task 3?

   c. Does it appear to you that the teams tend to be evenly matched?

5. For some rock concerts, audience members are chosen randomly to be checked for cameras, food, and other restricted items. Suppose that you observe the first 25 boys and 25 girls to enter a concert, and all 10 people chosen to be searched were male.

  a. How is this situation mathematically different from the one in the "On Your Own" on page 490?

  b. Describe a simulation model for this situation. What assumptions did you make in your model?

  c. Conduct at least 20 trials of your simulation. Record the results in a frequency table showing the number of boys selected.

  d. Is your conclusion different from your conclusion in the "On Your Own" task? Explain.

6. One of the sequences below is the result of actual flips of a coin. The other was written by a student trying to avoid doing the actual flips.

Sequence I

THHHHTTTTH	HHHTHHHHH	HHTTTHHTTH	HHHHTTTTTT
HHTHHTHHHT	TTHTTHHHHT	HTTTHTTTHH	TTTTHHHHHH
TTTHHTTHHH	THHHHHTTTT	THTTTHHTTH	TTHHTTTHHT
TTHHTHHTHH	TTTTTHHTHH	HHHTHTHTHT	HTHTTHHHTT
HHTHTHHHHH	HHHTTHTTHH	HHTTHTTTHH	TTTHHHTHHH

Sequence II

THTHTTTHTT	HTTHTHTTTH	TTHHHTHHTH	THTHTTTTHH
TTHHTTHHHT	HHHTTHHHTT	THHHTHHHHT	TTHTHTHHHH
THTTTHHHTH	HTHTTTHHTH	HHTHHHHTTH	THHTHHHTTT
HTHHHTHHTT	THHHTTTTHH	HTHTHHHHTH	TTHHTTTTHT
HTHTTHTHHT	THTTTHHTTT	HHHHTHTHHH	TTHHTHTTHH

  a. Which sequence do you think is the real one? Why did you select this one?

  b. Flip a coin 200 times, being sure each time that the coin spins many times in the air. Record the sequence of results.

  c. Does your sequence of actual coin flips look more like the first sequence or more like the second sequence?

Lesson 2

Estimating Expected Values and Probabilities

In trips to a grocery store, you may have noticed that boxes of cereal often include a surprise gift such as one of a set of toy characters from a current movie or one of a collection of stickers. Cheerios®, a popular breakfast cereal, once included one of seven magic tricks in each box.

Collect All 7 and Put On Your Own Magic Show!

MONEY CLIP TRICK
Make two clips magically join together!

MIND-READING TRICK
Guess the color your friend secretly picked!

VANISHING CARD TRICK
Make a card magically disappear!

MAGIC ROPE TRICK
Make the rope magically pass through solid tube!

DISAPPEARING COIN TRICK
Make a coin magically disappear and reappear!

SURPRISE 4'S
Turn two blank cards into two 4's!

MULTIPLYING COIN TRICK
Turn two coins into three!

Think About This Situation

This practice of cereal manufacturers raises interesting mathematical questions as well as marketing questions.

a Why do manufacturers include "surprises" in the packages with their product? How do you think they determine the number of different "gifts" to include?

b What is the least number of boxes you could buy and get all seven magic tricks?

c What is your prediction of the average number of boxes of Cheerios a person would have to buy to get all seven magic tricks?

d How could you simulate this situation?

INVESTIGATION ▶1 Simulation Using a Table of Random Digits

In the first lesson, you flipped a lot of coins to simulate situations. Calculators and computers can do this work for you. You can get strings of **random digits** either from your calculator or from a random digit table produced by a computer.

1. Examine this table of random digits between 0 and 9 inclusive generated by a computer.

2	4	8	0	3	1	8	6	5	6	4	2	0	0
7	6	8	6	3	0	5	6	8	2	5	0	7	5
0	9	5	8	1	7	3	0	9	9	8	7	7	7
0	2	6	8	6	2	5	5	4	1	5	9	8	0
4	1	2	9	0	8	6	7	0	3	3	8	2	1
1	5	8	0	9	5	7	3	5	6	5	0	2	3
9	7	6	2	5	9	2	6	3	5	0	3	1	3
2	1	0	9	6	0	1	8	5	5	2	2	6	6

   a. How many digits are in the table? About how many 6s would you expect to find? How many are there? Choose another digit and determine its frequency.

   b. About what percentage of digits in a large table of random digits from 0 to 9 will be even?

   c. About what percentage of the 1s in a large table of random digits from 0 to 9 will be followed by a 2?

   d. About what percentage of the digits in a large table of random digits from 0 to 9 will be followed by that same digit?

2. Refer to the table of random digits in Activity 1 when describing simulation models for the situations below.

   a. Explain how you could use the random digits to simulate whether a newborn baby is male or female. Describe a second way of using random digits for the same simulation.

   b. Explain how you could use the random digits to simulate tossing a die. (Disregarding particular digits won't affect the results.)

   c. Explain how you could use the random digits to simulate checking a box of Cheerios for which of seven magic tricks it contains. Does your plan require that you disregard some digits? Why or why not?

d. Explain how you could use the random digits to simulate spinning the spinner shown here.

e. Explain how you could use the random digits to simulate selecting three students at random out of a group of ten students. What does it mean to select students "at random"?

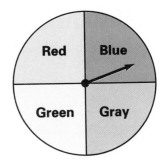

3. Obtain a table of one thousand random digits from your teacher or produce one using a calculator or computer.

   a. Explain how you could use this table of random digits to simulate the random selection of three students from a group of seven students. Perform one trial.

   b. Explain how you could use the table to simulate the experiment of rolling a die until you get a six. Perform one trial.

   c. Explain how you could use the table to simulate tossing two coins until both tosses are heads. Perform one trial.

4. Refer back to the description of magic tricks in Cheerios at the beginning of this lesson.

   a. If you buy one box of Cheerios, what is the probability that you will get a multiplying coin trick? To get your answer, what assumptions did you make about the tricks?

   b. Suppose you want to find the number of boxes of Cheerios you will have to buy before you get all seven magic tricks. Describe a simulation model for this. Describe how to conduct one trial using a table of random digits.

   c. Compare your group's simulation model with that of other groups. Then as a class, decide on a simulation model that all groups will use.

   d. Have each member of your group perform one trial using the agreed upon simulation model. How many "boxes" did each person in your group have to buy? Put these numbers in a frequency table such as the one below.

Number of Boxes to Get All 7 Tricks	Frequency	Number of Boxes to Get All 7 Tricks	Frequency
7		13	
8		14	
9		15	
10		16	
11		⋮	
12		**Total Number of Trials**	200

e. Complete a total of 200 trials of this simulation by sharing the work among the groups in your class. Complete a frequency table for the class data. Add additional rows if they are needed.

f. Find the average number of boxes purchased to obtain a complete set of the seven tricks. Use the frequency table and your calculator as needed. How does this estimate compare with the prediction you gave in Part c of the "Think About This Situation" on page 498?

g. If you didn't use simulation, how else could you estimate the average number of boxes needed? Which method is preferable? Why?

h. Make a histogram of the information in your frequency table. Explain how your histogram verifies that your answer to Part f is reasonable.

i. Describe the shape of your histogram. Is the shape of this distribution similar to any others you have constructed in this unit? If so, identify the situations related to those histograms.

Checkpoint

In this investigation, you have explored the properties of random digits and learned how to use them in designing a simulation.

ⓐ What is a table of random digits? How is it used in simulation?

ⓑ What are the advantages of using random digits in a simulation?

ⓒ To boost sales, tie-ins with popular movies are used by many types of manufacturers. As a tie-in to the animated film *The Hunchback of Notre Dame*, plastic figurines shaped like characters in the movie were distributed randomly in Cheerios boxes. Each box contained one of five different figurines. How could you simulate this situation efficiently?

Be prepared to share your group's descriptions and thinking with the entire class.

> ## On Your Own

A teacher notices that of the last 20 single-day absences in her class, 10 were on Friday. She wants to know if this can be attributed reasonably to chance or if she should look for another explanation.

a. Assuming that absences are equally likely to occur on each day of the school week, describe how to conduct one trial simulating the days of the week for 20 single-day absences using a table of random digits.

b. Conduct 10 trials. Place the results in a frequency table showing the number of absences that are on Friday.

c. Based on your simulation, what is your estimate of the probability of getting 10 or more absences out of 20 on Friday just by chance? What should the teacher conclude?

d. What is your estimate of the average number of absences on Friday, assuming that the 20 absences are equally likely to occur on each day of the week?

e. How could you get better estimates for Parts c and d?

INVESTIGATION ▶ 2 Simulation Using a Random Number Generator

A table of random digits is a convenient tool to use in conducting simulations. Playing cards and regular polyhedra models with numbered faces, such as dice, are sometimes useful. However, a more versatile tool is a calculator or computer software with a *random number generator*.

1. Investigate the nature of the numbers produced by the random number generator on your calculator or computer software. On some calculators, the generator is abbreviated "rand."

a. How many decimal places do the numbers usually have? Do some have one fewer place? If so, why?

b. Between what two whole numbers do all the random numbers lie?

c. Generate random numbers of the form "6 rand" (or "6 × rand"). Between what two whole numbers do all the random numbers lie?

d. Between what two whole numbers do the random numbers lie when "rand" is multiplied by 10? By 36? By 100?

e. Generate several numbers using the command "int 6 rand."

- What random numbers are generated by this function?
- What is the effect of the "int" function?

f. To simulate rolling a die, you need a random digit selected from the set 1, 2, 3, 4, 5, 6. In Part e, you generated random digits from the set 0, 1, 2, 3, 4, 5.

- How could you modify the command "int 6 rand" to produce the numbers on the faces of a die? Try it.
- Generate a list of 10 digits randomly selected from the set 1, 2, 3, 4, 5, 6.

2. Now explore how to generate random digits from other sets of numbers.

a. What random digits are generated by the command "int 10 rand + 1"?

b. Modify the command in Part a to generate random digits from the set:

- 1 to 23 inclusive
- 1 to 100 inclusive
- 1 to 52 inclusive
- 0 to 6 inclusive

Test your modified commands.

3. Next, consider some of the possible contexts in which a random number generator might be useful.

a. Explain how you could use the random number generator to simulate the flips of a fair coin.

b. Explain how you could use the random number generator to select three students at random from a group of seven students. Perform one trial of this simulation.

c. Explain how you could use the random number generator to select six students at random from a group of 50 students. Perform one trial of this simulation.

d. Explain how you could use the random number generator to simulate the experiment of rolling a die until you get a six. Perform one trial of this simulation.

e. Explain how you could use the random number generator to simulate checking a box of Cheerios for which magic trick it contains. Perform one trial of this simulation.

f. Explain how you could use the random number generator to simulate the experiment of drawing a card from a well-shuffled deck of 52 playing cards and checking if it is an ace. Perform one trial of this simulation.

4. Software for computers and calculators has been developed to help you quickly conduct many trials of a simulation. This activity will illustrate how you can use such software to implement your simulation model for the Cheerios problem. The calculator software *Collections* (COLLECT) is an example.

a. Use such software to conduct 25 trials of collecting the seven magic tricks.

b. Study a histogram of the results of your 25 trials. What do the bars in the display represent?

c. Compare your histogram with those of other members of your group. Are they the same? If not, what accounts for the differences?

d. With the COLLECT program, you can use the arrow keys to explore the histogram. What do the "min =," "max =," and "n =" mean? (Remember that the number between two bars belongs in the bar to the right of the number.)

e. What is your estimate of the average number of boxes of cereal you would have to buy before you obtained all 7 tricks? (With COLLECT, press ENTER from the histogram screen. This information will be displayed.)

f. Have each member of your group use the software to conduct different numbers of trials (for example, 50, 75, or 99). Record and compare your results.

g. Suppose the cereal manufacturer modified its marketing scheme. It randomly packaged one of six differently colored pens in each box of cereal. Use the software to help you estimate the average number of boxes of Cheerios you would have to purchase in order to get a complete set of the six pens.

Checkpoint

When using a calculator or computer to generate random numbers,

a how can you control the size of these numbers?

b how can you ensure they will be integers?

Be prepared to share your group's thinking with the class.

On Your Own

Lynn is taking a 10-question multiple-choice test. Each question offers four possible choices for the answer. She didn't study and doesn't even have a reasonable guess on any of the questions. For each question, Lynn selects one of the four possible answers at random.

a. Describe how to conduct one trial simulating a 10-question test. Use the random number generator on your calculator or a computer.

b. Conduct 20 trials and place the results in a frequency table.

c. What is your estimate of the probability that Lynn will pass the test with a score of 70% or more?

MORE

Modeling • Organizing • Reflecting • Extending

Modeling

1. Cracker Jack®, the caramel corn-peanut snack, traditionally gives a prize in each box. At one time, it offered five different prizes.

   a. Describe how to conduct one trial of a simulation for estimating the number of boxes of Cracker Jack® you would have to buy before you get all five prizes. What assumptions are you making?

   b. Conduct five trials. Copy the following frequency table, which has the results from 195 trials. Add the results from your five trials to the table.

Number of Boxes Purchased	Frequency
5	6
6	15
7	23
8	18
9	21
10	18
11	15
12	15
13	11
14	11
15	10

Number of Boxes Purchased	Frequency
16	6
17	6
18	8
19	1
20	2
21	4
22	0
23	0
24	1
25	2
26	0

Number of Boxes Purchased	Frequency
27	1
28	0
29	0
30	0
31	0
32	0
33	0
34	0
35	0
36	1

c. What is the average number of boxes purchased in the 200 trials to get all five prizes?

d. Make a histogram of the (*number of boxes purchased, frequency*) data. Describe the shape of the histogram.

e. Explain how you can use the histogram to verify the average you found in Part c.

2. Toni doesn't have a key ring and so just drops her keys into the bottom of her backpack. Her four keys—a house key, a car key, a locker key, and a key to her bicycle lock—are all about the same size.

a. If she reaches into her backpack and grabs the first key she touches, what is the probability it is her car key?

If the key drawn is not her car key, she holds onto it. Then, without looking, she reaches into her backpack for a second key. If that key is not her car key, she holds on to both keys drawn and reaches in for a third key.

b. Do the chances that Toni will grab her car key increase, decrease, or remain the same on each grab? Explain your reasoning.

c. Describe how to conduct one trial of a simulation to estimate the probability that Toni has to grab all four keys before she gets her car key. Use a table of random digits, a calculator, or computer software to conduct the trial.

d. Conduct 10 trials. Copy the frequency table below and add your results so that there is a total of 1,000 trials.

Number of Keys Toni Needs to Grab to Get Her Car Key	Frequency
1	253
2	250
3	233
4	254
Total Number of Trials	

e. What is your estimate of the probability that Toni has to grab all four keys before she gets her car key?

f. From the frequency table, it appears that the numbers of keys needed are equally likely. Explain why this is the case.

3. One cereal promotion offered one of four items for little girls: a comb, a mirror, a barrette, or a bracelet. The catch was that each item came in four colors: yellow, pink, blue, and lavender. Suppose that María wants a complete set, all in lavender.

a. Describe how to use a table of random digits or a random number generator to conduct one trial simulating the number of boxes María's family will have to buy. What assumptions are you making?

b. Conduct five trials. Add your results to a copy of the stem-and-leaf plot at the right. There will be a total of 50 trials.

c. What is your estimate of the probability that more than 15 boxes of cereal must be purchased to get a complete set all in lavender?

1	04466889
2	012234466778899
3	133344578
4	01456
5	0016788
6	6

2|4 represents 24

d. What is your estimate of the average number of boxes that must be purchased?

e. Will the median number of boxes be more or less than the average? Find the median to check your prediction.

4. Suppose Whitney also knows about the promotion described in Modeling Task 3. She wants a complete set in one color, but any color is okay.

   a. Describe how to use a table of random digits or a random number generator to conduct one trial simulating the number of boxes Whitney's family will have to buy. What assumptions are you making?

   b. Conduct 10 trials. Combine your results with those from other members of your group. Place the results in a frequency table.

   c. What is your estimate of the probability that more than 15 boxes of cereal must be purchased to get a complete set all in the same color?

   d. What is your estimate of the average number of boxes that must be purchased?

5. The card below shows a McDonald's-Atari Asteroids® scratch-off game. All of the asteroids were originally covered. The instructions say:

   Start anywhere. Rub off silver asteroids one at a time. Match 2 identical prizes BEFORE a "ZAP" appears and win that prize. There is only one winning match per card.

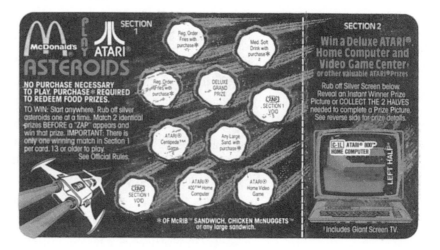

   a. Describe how to conduct one trial of a simulation to determine the probability of winning the prize in the scratch-off game. Share your instructions for how to perform a trial with your group. Modify the sets of instructions to reach a consensus on the model.

   b. Exchange your instructions with another group. Conduct 20 trials using the other group's model.

   c. Combine the results from all of the groups in your class and estimate the probability of winning the prize.

Organizing

1. Make a box plot of the data from the 200 trials simulating the Cheerios problem in Part e of Activity 4 (page 501).

   a. How does the shape of the box plot reflect the shape of the histogram you prepared for Part h of that activity? In which direction are the plots skewed?

   b. Complete this sentence about the distribution of boxes purchased to get all 7 magic tricks: *Half of the time, a person who wanted all seven magic tricks would have to buy at least _____ boxes of Cheerios.* Does this statement refer to the average number of boxes or the median number of boxes?

2. Make a box plot of the distribution in one of the Modeling tasks you completed. Compare this box plot to the one you made in Organizing Task 1. Do the two plots have the same basic shape? Explain.

3. Suppose different cereal companies have packaged in their boxes the number of different prizes given below. Carry out simulations to estimate the average number of boxes that would have to be purchased to get all of the prizes in each case. (The software you used for Activity 4, Investigation 2 on page 504 would be helpful.)

   a. 2 different prizes   **b.** 3 different prizes

   c. 5 different prizes   **d.** 8 different prizes

   e. 20 different prizes

   f. Create a table with column headings "Number of Prizes" and "Average Number of Boxes Purchased." Fill the table with your responses to Parts e and g of Activity 4, Investigation 2 (page 504) and your data for this task (Parts a–e above). Make a scatterplot of the (*number of prizes, average number of boxes purchased*) data.

   ■ Does a linear model fit these data reasonably well? If so, find the regression line. If not, describe the shape of the graph.

   ■ About how many additional boxes must be purchased if one more type of prize is added to the boxes?

4. Toni's key selection problem (Modeling Task 2) is one that depends on *order*—the order in which she chooses the keys. One way to model the problem would be to list all the possible orders in which the keys could be selected. An ordering of a set of objects is called a **permutation** of the objects. For example, the permutations of the letters A, B, and C are:

   ABC   ACB   BAC   BCA   CAB   CBA

   a. In what special sequence are the permutations above listed?

   b. List all of the permutations of the letters A and B. How many are there?

c. List all of the permutations of the letters A, B, C, and D. How many are there?

d. Look for a pattern relating the number of permutations to the number of letters.

e. How many permutations do you think there are of the letters A, B, C, D, and E? Check your conjecture by listing all of the permutations or by using the permutations option on your calculator or computer software. (For most calculators, you need to enter "5 nPr 5". This means the number of permutations of 5 objects taken 5 at a time.)

f. How many permutations are there of Toni's four keys? What is the probability that all four keys have to be drawn before Toni gets her car key?

Reflecting

1. You have conducted simulations by coin flipping and by using random digits from a table, calculator, or computer.

   a. Which of these simulation tools is easiest for you to understand?

   b. Which tool is the most flexible in simulating a variety of situations?

   c. Which tool do you find easiest to use?

2. A deck of playing cards can be used to simulate some situations.

   a. How could you use cards to conduct one trial in a simulation of collecting Cheerios tricks?

   b. How could you use cards to generate a table of random digits?

3. Consider the calculator command "int 12 rand + 1".

   a. What random numbers are generated by this command?

   b. Give an example of a real-life situation which might lead to a simulation model involving use of this calculator or computer command.

4. Do you think it is faster to do a simulation using a table of random digits or using a calculator to generate random digits? Design and carry out an experiment to answer this question. Write a brief summary describing your experiment and findings.

5. The COLLECT program will conduct up to 99 trials of certain simulations. Suppose you need 200 trials simulating a collecting problem. How could you use the program to help you?

Extending

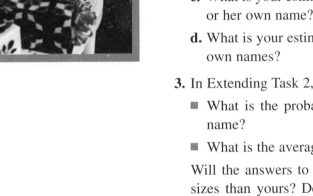

1. There is some evidence that either there were more of some of the Cheerios magic tricks than others, or the magic tricks weren't put into the boxes randomly. (This might happen, for example, if all of the boxes containing the disappearing coin trick were sent to California.) Suppose you are more likely to get some magic tricks than others. Will the average number of boxes purchased in order to get all seven magic tricks be more, the same, or less than if each trick is equally likely? Design and carry out a simulation to answer this question.

2. Suppose your class is having a holiday gift exchange. The name of each student is written on a card. The cards are placed in a hat, mixed up, and each student draws one. One of your classmates wonders how many students will draw their own names.

   a. Describe one trial of a simulation of this situation. Use cards or a table of random digits.

   b. Conduct 50 trials. Place the results in a table that shows the number of students who got their own name.

   c. What is your estimate of the probability that at least one student will get his or her own name?

   d. What is your estimate of the average number of students who will get their own names?

3. In Extending Task 2, you answered these two questions:

   ■ What is the probability that at least one student will get his or her own name?

   ■ What is the average number of students who will get their own names?

   Will the answers to these two questions change with larger or smaller class sizes than yours? Design and carry out simulations to test your prediction. Write a summary of your findings.

4. One type of CD player will hold five CDs. The player can be set so that it selects a CD at random and plays a song on that CD. It then continues selecting CDs at random from all five CDs and playing songs. Describe one trial of a simulation to estimate the probability that there is at least one song from each of the five CDs among the first ten songs played.

5. Imagine 10 people in a room.

    a. What is your estimate of the probability that two of them have the same birthday?

    b. Describe a simulation model to determine the probability that two people out of a group of 10 people have the same birthday. What assumptions are you making?

    c. Conduct 100 trials, sharing the work with other members of your class.

    d. From your simulation, what is your estimate of the probability? Are you surprised?

    e. Investigate the probability that two people will have the same birthday if there are 20 people in the room. Repeat for cases when there are 30, 40, 60, and 80 people in the room.

    f. People often make too much of coincidences. Write an explanation of the following Drabble cartoon for a child who doesn't "get it."

DRABBLE reprinted by permission of United Feature Syndicate, Inc.

Simulation and the Law of Large Numbers

Lesson 3

In 1985, Major League Baseball switched from a best-of-five league playoff series to a best-of-seven league playoff series. In a best-of-five series, the team that first wins three games wins the series.

Think About This Situation

The winners of the league playoffs represent the National and American Leagues in the World Series. This event is also a best-of-seven series.

a How many games do you have to win to be victorious in a best-of-seven series?

b Why do you think Major League Baseball went from a five-game to a seven-game championship series?

INVESTIGATION ▶ 1 How Many Games Should You Play?

In this investigation, you will explore playoff series of various lengths. You will begin with the simplified situation in which the probability a team wins stays the same from game to game.

1. The Cyclones are playing the Hornets for the softball championship. Based on their history, the Cyclones have a 60% chance of beating the Hornets in any one game.

   a. Suppose the championship series were only one game long. What is the probability that the better team (Cyclones) would win?

   b. Suppose the Cyclones and the Hornets were to play 10 games. About how many games would you expect the Cyclones to win? The Hornets? Explain your reasoning.

c. Describe how you would design one trial of a simulation model for a one-game series. Use a table of random digits or a random number generator. How will you split the numbers so that the Cyclones have a 60% chance of winning and the Hornets have a 40% chance of winning?

d. Why isn't it appropriate to use a coin flip in your model to determine which team wins a game?

2. Should an even number of games be used for a playoff series? Explain your reasoning.

3. Suppose the Cyclones and the Hornets play a three-game series. A three-game series ends after one team wins two games.

a. Describe how to conduct one trial of a simulation model for a three-game series.

b. Describe how you could use 200 trials of your simulation model to estimate the probability that the better team (Cyclones) will win a best-of-three series.

c. Conduct 200 trials. Share the work among the groups in your class. Record your data in a frequency table like the one below.

Number of Games Won by the Cyclones in a Best-of-Three Series	Frequency
0	
1	
2	
Total Number of Trials	200

d. Which rows represent a win of the championship series by the Cyclones? By the Hornets?

e. What is your estimate of the probability that the Cyclones will win a best-of-three series?

4. Next, explore the idea of a five-game series between the same two teams.

a. Describe how to conduct one trial of a simulation model for a best-of-five series.

b. Describe how you could use your simulation model to estimate the probability that the Cyclones will win a best-of-five series.

c. Conduct 200 trials by sharing the work among the groups in your class. Record your data in a frequency table where the number of games won by the Cyclones goes from 0 to 3. Why do you only need numbers up to 3?

d. What is your estimate of the probability that the Cyclones will win a best-of-five series?

5. Now explore the idea of a seven-game series for the same teams.

   a. Describe how to conduct one trial of a simulation for a seven-game series.

   b. Describe how you could use your simulation to estimate the probability that the Cyclones will win a best-of-seven series.

   c. Conduct 200 trials. Share the work among the groups in your class. Place your results in a frequency table showing the number of games won by the Cyclones.

   d. What is your estimate of the probability that the Cyclones will win a best-of-seven series?

6. Complete the table below using the results from Activity 1 Part a and Activities 3, 4, and 5.

Type of Series	Estimate of Probability the Cyclones Win
One game	
Best-of-three	
Best-of-five	
Best-of-seven	

   a. What pattern do you observe in the table? Which team, the Cyclones or the Hornets, would have the better chance to win a best-of-nine series? Estimate the probability of their winning.

   b. Improve your estimate in Part a by carrying out a simulation.

   c. From a mathematical point of view, why do you now think that Major League Baseball went from a five-game to a seven-game championship series?

7. Make histograms of your frequency tables from Activities 3–5. How are these histograms alike? How are they different?

8. Tennis players have two chances to make a legal serve. Monica makes about 50% of her *first* serves. If she has to try a *second* serve, Monica makes about 80% of those.

   a. Describe how to use a table of random digits to conduct one trial simulating this situation.

   b. Describe how to use a random number generator to conduct one trial simulating this situation.

c. Conduct one trial of your simulation model and record the result in a frequency table. Your frequency table should have three rows: Makes First Serve, Misses First Serve and Makes Second Serve, and Double-Faults (misses both serves).

d. Conduct 50 trials.

e. Estimate the probability that Monica double-faults.

In the previous situations, the percentages of *success*—winning a game or making a serve—were multiples of 10. In those cases, you may have used single digits from your random digits table or from your calculator. Often, as in the next two situations, the percentages are not as "nice."

9. A survey of high school seniors found that 29% had seen a movie in the previous two weeks.

a. Describe how to conduct one trial of a simulation to estimate the number of recent movie-goers in a randomly selected group of 20 seniors.

b. Conduct 10 trials, placing the results on a number line plot that shows the number of recent movie-goers in each group of 20 randomly selected students.

10. In the 50-year history of National Basketball Association finals, the home team has won about 71% of the games. Suppose that the Los Angeles Lakers are playing the Philadelphia 76ers in the NBA finals. The two teams are equally good, except for this home team advantage. The finals are a best-of-seven series. The first two games will be played in Philadelphia, the next three (if needed) in Los Angeles, and the final two (if needed) in Philadelphia.

a. What is the probability that the 76ers will win a game if it is at home? What is the probability that the 76ers will win a game if it is played in Los Angeles? What is the probability the 76ers will win the first game of the series? The second game? The third game? The fourth game? The fifth game? The sixth game? The seventh game?

b. Describe how to conduct one trial to simulate a finals series.

c. Conduct 200 trials by sharing the work among the groups in your class. Place the results in a frequency table like the one on the next page.

Number of Games Won by the 76ers	Frequency
0	
1	
2	
3	
4	
Total Number of Trials	200

    d. What is your estimate of the probability that the 76ers win the finals?

    e. Suppose that, to cut travel costs, the NBA schedules three games in Los Angeles followed by four in Philadelphia.

        ■ Design a simulation to determine the probability that the 76ers win the finals in this situation.

        ■ Conduct 200 trials by sharing the work with other groups.

        ■ What is your estimate of the probability that the 76ers win this series?

    f. Compare the probabilities in Parts d and e. What is your conclusion?

Checkpoint

In this investigation, you learned how to design simulations for situations in which the outcomes are not equally likely. You also explored why it is better to conduct a large number of trials of a simulation.

a In playoff series, what is the advantage of a longer series over a shorter one?

b How can random numbers be used in simulations when the two outcomes are not equally likely?

c Sheila has a 55% chance of winning a table tennis game against Bobby. Describe a simulation model for estimating the probability that Sheila wins a best-of-nine series of table tennis games against Bobby. Should Bobby prefer a best-of-three series?

Be prepared to share your group's thinking and simulation model with the class.

The simulations in this unit were based on a variation of the **Law of Large Numbers**. The Law of Large Numbers says, for example, that if you roll a die more and more times, the proportion of fives tends to get closer to $\frac{1}{6}$. The

Cyclones have a 60% chance of winning each game. For a very long series, the Law of Large Numbers says that the percentage of games the Cyclones win tends to be close to 60%. (So the Cyclones are almost sure to win a very long series.)

On Your Own

Recall that in Major League Baseball, the World Series is a best-of-seven-games series. In Modeling Task 3 on page 492, you estimated the probability that the World Series will go seven games if the teams are equally matched. That probability is actually 0.3125.

a. Suppose the two teams aren't evenly matched. Do you think the World Series is more likely to go seven games or less likely to go seven games than if the teams are evenly matched? Why?

b. Suppose that the teams are not evenly matched and that the American League team has a 70% chance of winning each game. Describe how to conduct one trial of a best-of-seven series for this situation.

c. Conduct 5 trials. Add your results to a copy of the frequency table below so that there is a total of 100 trials.

Number of Games Needed in the Series	Frequency
4	24
5	30
6	24
7	17
Total Number of Trials	

d. What is your estimate of the probability that the series will go seven games in this case? Is this probability of a seven-game series more or less than when the teams are evenly matched?

MORE

Modeling • Organizing • Reflecting • Extending

Modeling

1. About 60% of high school transcripts in the United States show that the student has taken a chemistry or physics course.

   a. Describe how to conduct one trial of a simulation to estimate the number of students who took a high school chemistry or physics course in a randomly selected group of 20 high school graduates.

   b. Conduct 5 trials. Add your results to a copy of the frequency table below so there is a total of 400 trials.

Number Who Took Chemistry or Physics	Percentage Who Took Chemistry or Physics	Frequency
6	30	3
7	35	2
8	40	17
9	45	39
10	50	51
11	55	53
12	60	80
13	65	66
14	70	39
15	75	29
16	80	13
17	85	3

Total Number of Trials

   c. Make a histogram of these results. Place "Percentage Who Took Chemistry or Physics" on the horizontal axis.

   d. When 20 students are randomly selected, what is your estimate of the probability that fewer than half took chemistry or physics in high school?

2. The following histogram shows the result of 400 trials of a simulation. The situation modeled is the same as in Task 1 except that 80 students were randomly selected.

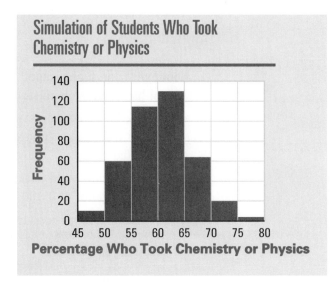

Simulation of Students Who Took Chemistry or Physics

a. Estimate the probability that fewer than half of 80 randomly selected students took chemistry or physics.

b. Why do you think the probabilities in Part d of Modeling Task 1 and Part a above are different?

c. Describe how the histograms from Modeling Tasks 1 and 2 are different. Why would you expect this to be the case? How do these histograms illustrate the Law of Large Numbers?

d. Estimate the probability that fewer than half of a random selection of 320 students took chemistry or physics in high school. Explain your reasoning.

3. Several years ago, a survey found that 25% of American pet owners carry pictures of their pets in their wallets.

a. Assume this percentage is true. Describe how to simulate determining if a randomly chosen American pet owner carries a picture of his or her pet. Use a calculator, computer, or table of random digits in your simulation.

b. Describe how to conduct one trial that models how many people in a random sample of 20 American pet owners carry pictures of their pets.

c. Perform 10 trials of your simulation. Add your results to a copy of the following frequency table so that there is a total of 100 trials.

Number of People with a Picture of Their Pet	Frequency
0	0
1	2
2	6
3	12
4	17
5	18

Number of People with a Picture of Their Pet	Frequency
6	15
7	10
8	5
9	3
10	1
11	1
Total Number of Trials	

d. Take a survey of 20 American pet owners to see how many carry pictures of their pets in their wallets. Do you have any reason to doubt the reported figure of 25%?

4. A library has 25 computers which must be reserved in advance. The library estimates that people will decide independently whether or not to show up for their reserved time and that for each person there is an 80% chance that he or she will show up.

   a. If the library takes 30 reservations for each time period, how many are expected to actually show up? Is it possible that all 30 will show up?

   b. How many reservations should the library take so that 25 are expected to show up? Is it possible that if the library takes this many reservations, more than 25 people will show up?

   c. Suppose the library decides to take 30 reservations. Describe one trial in a simulation of this situation.

   d. Conduct 5 trials. Add your results to a copy of the frequency table below so that there is a total of 55 trials.

Number of People Who Show Up	Frequency
19	0
20	1
21	2
22	3
23	5
24	8

Number of People Who Show Up	Frequency
25	9
26	9
27	7
28	4
29	2
30	0
Total Number of Trials	

e. Make a histogram of the results of this simulation. Describe the shape of your histogram.

f. Suppose the library takes 30 reservations. Based on your simulation, estimate the probability that more than 25 people show up. Is 30 a reasonable number of reservations to take? Should the library take fewer or could it get away with taking more?

g. What assumption made in this simulation is different from the real-life situation being modeled?

Organizing

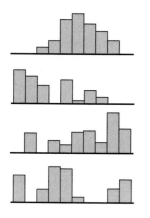

1. Examine all of the histograms that you have made in this unit. Group them according to their general shape. What observations can you make about the shapes and kinds of probability situations associated with them? For example, which kinds of situations give distributions that have line symmetry?

2. Explain why it is important to conduct as many trials as you can in a simulation.

3. Suppose you want to be at least 95% sure that a probability estimate from a simulation is within a certain *margin of error* of the actual probability. The formula gives the number of trials n needed in order to estimate within a margin of error E.

$$n = \frac{1}{E^2}$$

For example, suppose you want to estimate the probability that a bent coin comes up heads. You want to be at least 95% sure your estimated probability is within $E = 0.1$ (10%) of the actual probability. You need to perform trials of the simulation.

$$n = \frac{1}{(0.1)^2} = \frac{1}{0.01} = 100$$

Being "at least 95% sure" means that out of every 100 simulations you perform, you expect that in 95 or more of them your estimated probability is within E of the actual probability.

a. Suppose you want to estimate a probability in a simulation and be at least 95% sure that your estimated probability is within 5% of the actual probability. How many trials should you perform?

b. In some of the simulations in this unit, you performed 200 trials. What is the margin of error associated with 200 trials of a simulation?

c. What is the margin of error associated with 1,000 trials of a simulation?

4. Use your graphing calculator or computer software to investigate the graph of the *number-of-trials* function given in Organizing Task 2.

   a. Should x or y represent the number of trials? What does the other variable represent?

   b. Set Xmin = 0.01 and Xmax = 0.15. What are reasonable choices for Xscl, Ymin, Ymax, and Yscl?

   c. Describe the shape of the graph of this function, including any symmetry.

   d. Use the trace function to estimate the number of trials needed for a margin of error of 2%, 3%, and 8%.

   e. How could you answer Part d using the table-building capability of your calculator or computer software?

5. In this task, you will investigate the number of trials of a simulation needed to cut the margin of error in half.

   a. In Organizing Task 3, you saw that with 100 trials, there is a margin of error of 10%. How many trials would you need to cut this margin of error in half?

   b. With 625 trials there is a margin of error of 4%. How many trials would you need to cut this margin of error in half?

   c. In general, to cut a margin of error in half, how must the number of trials change?

Reflecting

1. The Law of Large Numbers tells you that if you flip a coin repeatedly, the percentage of heads tends to get closer to 50%.

   a. Does the table below illustrate the Law of Large Numbers? Why or why not?

   b. Do you see a surprising result in the table? Explain why you find it surprising.

Number of Flips	Number of Heads	Percentage of Heads	Expected Number of Heads	Excess Heads
10	6	60%	5	1
100	56	56%	50	6
1,000	524	52.4%	500	24
10,000	5,140	51.4%	5,000	140

2. Suppose your class is planning a checkers tournament and you are sure you are the best player in the class. What kind of tournament rules would you propose? Explain your reasoning.

3. One simulation method, called the *Monte Carlo method*, was developed during World War II at Los Alamos, New Mexico, to solve problems that arose in the design of atomic reactors. The method is once again an area of active research, this time with applications to ecology, genetics, sociology, political science, and epidemiology (the study of the prevalence and spread of disease). Investigate the history of the Monte Carlo method and prepare a brief report.

4. You have now used simulation extensively to estimate answers to problems involving chance.

   a. What advantages do you see in using simulation?

   b. What disadvantages do you see in using simulation?

5. The mathematician Pierre Simon, Marquis de Laplace (1749–1827), once said:

> *"It is remarkable that a science which began with the consideration of games of chance should have become the most important object of human knowledge. ... The most important questions of life are, for the most part, really only problems of probability."*

   a. What do you think Laplace meant by the last statement?

   b. Do you agree with him? Why or why not?

Pierre Simon, Marquis de Laplace

Extending

1. Acceptance sampling is one method that industry uses to control the quality of the parts they use. For example, a recording company buys blank cassette tapes from a supplier. To ensure the quality of these tapes, the recording company examines a sample of the tapes in each shipment. The company buys the shipment only if 5% or fewer of the tapes in the sample are defective. Assume that 10% of the tapes actually are defective.

"By a small sample we may judge the whole piece."
Miguel de Cervantes

a. Suppose the recording company examines a sample of 20 tapes from each shipment. Design and carry out a simulation of this situation. What is your estimate of the probability that the shipment will be accepted?

b. Suppose the recording company examines a sample of 100 tapes from each shipment. Design and carry out a simulation of this situation. What is your estimate of the probability that the shipment will be accepted?

2. Laboratory tests indicate that when planted properly, 6% of a particular type of seed fail to germinate. This means that out of every 100 seeds planted according to instructions, on the average 6 do not sprout. The laboratory has been developing a new variety of the seed in which only 1% fail to germinate. Suppose that in an experiment, 10 seeds of each of the two types are planted properly.

a. For each type of seed, make a prediction of the probability that *at least one* seed out of the 10 will fail to germinate.

b. Design and carry out a simulation to estimate the chance that if 10 of the seeds with the 6% rate are planted, *at least one* will fail to germinate.

c. Design and carry out a simulation to estimate the chance that if 10 of the new variety of the seed are planted, *at least one* will fail to germinate.

d. Compare the estimates from your simulations to your predictions in Part a. What have you learned?

3. The Martingale is an old gambling system. At first glance, it looks like a winner. Here's how it would work for a player betting on red in roulette. The roulette wheel has 38 spaces; 18 of these spaces are red. On the first spin of the wheel, Tom bets $1. If red appears, he collects $2 and leaves. If he loses, he bets $2 on red on the second spin of the wheel. If red appears,  he collects $4 and leaves. If he loses, he bets $4 on red on the third spin of the wheel, and so on. Tom keeps doubling his bet until he wins.

a. If Tom wins on his first try, how much money will he be ahead? If he wins on his third try, how much money will he be ahead? If Tom finally wins on his tenth try, how much money will he be ahead?

b. From a gambler's point of view, what are some flaws in this system?

c. Design and carry out a simulation to test the Martingale system on roulette.

d. Write a report on this system for a newspaper.

Looking Back

In this unit, you have used simulation models to help solve problems involving chance. An important feature of all your models was the use of coins, dice, or random numbers to produce random outcomes. In each case, the outcomes had the same mathematical characteristics as those in the original problem.

Almost any problem involving probability or an expected value can be solved using simulation models. This final lesson of the unit provides three more such problems to help you pull together the ideas you have developed and increase your confidence in using simulation methods.

1. About 10% of the adult population of the United States are African-American.

   a. Jurors are selected for duty in their city or town. Consider a city which has an African-American population representative of the U.S. population. Design a simulation model to determine the probability that a randomly selected jury of 12 people would have no African-American members. Write instructions for performing one trial of your simulation. Exchange instructions with another group.

   b. Do the other group's instructions model the situation well? If necessary, modify the instructions and then conduct 5 trials. Add your results to a copy of the following frequency table so that there is a total of 200 trials.

Number of African-Americans on the Jury	Frequency		Number of African-Americans on the Jury	Frequency
0	56		3	16
1	73		4	4
2	45		5	1

Total Number of Trials

   c. Make a histogram of the results in your frequency table.

   d. What is your estimate of the probability that a randomly selected jury of 12 people would have no African-American members?

e. A *grand jury* decides whether there is enough evidence against a person to bring him or her to trial. A grand jury generally consists of 24 people. Do you think the probability that a randomly selected grand jury would have no African-American members is more, less, or the same as your answer to Part d? Why?

f. Describe a simulation to find the probability that a randomly selected grand jury would have no African-American members. Conduct 5 trials of your simulation model.

2. This roller coaster has 7 cars. Ranjana stands in a long line to get on the ride. When she gets to the front, she is directed by the attendant to the next empty car. No one has any choice of cars, but must take the next empty one in the coaster. Ranjana goes through the line 10 times. She likes to sit in the front car.

a. Each time she goes through the line, what is the probability Ranjana will sit in the front car? Do you think Ranjana has a good chance of sitting in the front car at least once in her 10 rides? Explain your reasoning.

b. Describe how to conduct one trial simulating this situation. Use your calculator or computer or a table of random digits.

c. Perform 15 trials. Place the results in a frequency table that lists the number of times out of the 10 rides that Ranjana sits in the front car.

d. From your simulation, what is your estimate of the probability that Ranjana will sit in the front car at least once?

e. How could you get a better estimate for Part d?

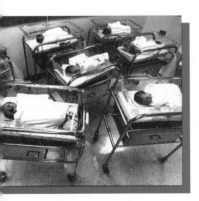

3. The chance that a newborn baby will be a girl is about $\frac{1}{2}$. Suppose that in one large hospital, 20 babies are born each day. In a smaller hospital nearby, 5 babies are born each day.

a. Do you think the size of the hospital affects the number of days in which 60% or more of the babies born are girls? If so, do you think this would happen more often in the large hospital or the small one? Explain your reasoning.

b. Describe how to conduct one trial of a simulation model to find the number of girls born on one day in the large hospital. Describe a similar model for the small hospital.

c. Conduct 20 trials of a simulation for both the large and the small hospital to test the conjecture you made in Part a. Be sure to make a histogram of your results.

d. Are the results of your simulations different from your conjecture? What should you conclude?

e. Shown below are responses to the question in Part a by a group of college undergraduates.

Small Hospital	Large Hospital	No Difference
17	15	48

Why do you think so many undergraduates believed the size of the hospital would not make a difference?

f. How is this problem related to your work in the previous lesson on the best length for a playoff series?

Checkpoint

In this unit, you learned how to model situations involving chance using simulations.

a Summarize the steps involved in using a simulation model to solve a problem involving chance.

b Will a simulation give you an "exact" answer? Explain your reasoning.

c What does the Law of Large Numbers say about how many trials should be done in a simulation?

d A letter to the *Washington Post* on May 11, 1993, suggested that China has more boys born than girls because if the first child is a boy then the parents tend to stop having children. Based on your work in this unit, do you think this is likely to be the case? Write a response to the author of this letter explaining your reasoning.

Be prepared to share your summary and explanations with the entire class.

On Your Own

Write, in outline form, a summary of the important mathematical concepts and methods developed in this unit. Organize your summary so that it can be used as a quick reference in your future work.

Looking Back
at Course 1

Planning a Benefits Carnival

In this course, you have built useful mathematical models—including linear, exponential, geometric, simulation, and graph models. You have used these models to solve important problems in many different settings. You have investigated patterns in data, change, chance, and shape. And you have learned how to make sense of situations by representing them in differ-

ent ways using physical representations, words, graphs, tables, and symbols. In this Capstone, you will pull together many of the important concepts, techniques, and models that you have learned and use them to analyze one big project.

Think About This Situation

Many schools organize fund-raising events to raise money for improving their programs. The event might be a dance, a bike-a-thon, a book sale—anything that gets the community involved in raising money for the school. One event that is common for elementary schools is a benefits carnival. Suppose that a local elementary school is considering such a carnival. Your class has offered to plan the event and prepare a full report for the school's principal. (In return, your class will get part of the proceeds to use for yourselves!)

ⓐ Make a list of all the things that need to be done to plan, carry out, and clean up following such a carnival.

ⓑ Make a list of the kinds of booths or activities that might be good to have at the carnival. Think of as many as you can.

INVESTIGATION 1 ▶ Lots of Math

Mathematics can be used in many different ways to help you organize the carnival. Think about the mathematics you have studied in each of the units in this course. The units are listed below. As a group, brainstorm and then write two ways the mathematics in each unit could be used in the carnival project.

1. *Patterns in Data*
2. *Patterns of Change*
3. *Linear Models*
4. *Graph Models*
5. *Patterns in Space and Visualization*
6. *Exponential Models*
7. *Simulation Models*

Checkpoint

Different groups probably identified different ways in which mathematical ideas in Units 1–7 could be used in the carnival project.

a For each unit, compare and discuss the ideas from different groups.

b Are there any big mathematical ideas or topics from this course that have not been applied to the carnival project? If so, is there any way they could be applied?

Be prepared to share your thinking with the whole class.

At the end of this Capstone, you will write an individual report for the principal of the elementary school. In this report, you need to explain how you used mathematics to help plan the carnival. To assist in the preparation of the reports, your group will complete three of the following investigations. Each group will present an oral report to the class on one of them. (Guidelines for the group report are given on page 544.)

As a group, examine Investigations 2 through 8 and choose three to complete. Confirm your choices with your teacher, and then start investigating!

INVESTIGATION ▶ 2 Careful Planning

Careful planning is necessary to make the carnival a success. An important part of planning is identifying and scheduling all the tasks that need to be done.

1. Some of the tasks and task times for the carnival project might be building the booths (5 days), laying out the floor plan (2 days), designing a carnival logo (1 day), publicity (10 days), and finding a location (2 days). Write at least two more tasks that will need to be done. Estimate the times needed to complete these tasks.

2. Use the tasks from Activity 1 to do the following:

   a. Construct a prerequisite table showing the tasks, the task times, and the immediate prerequisites.

   b. Draw a project digraph.

   c. Find the critical tasks and the earliest finish time for the whole project.

   d. Set up a schedule for completing all the tasks. Your schedule should show earliest start times, earliest finish times, and slack times.

3. Committees must be formed to work on each of the tasks. Since some students will be on more than one committee, it is impossible for all the committees to meet at the same time. Assume there are five committees, each of which has a member in common with at least two other committees.

a. Construct a vertex-edge graph model that shows the five committees and which committees share members. (You decide which committees share members.)

b. What is the fewest number of meeting times needed so that all committees can meet? Explain how you can use the graph model from Part a to answer this question.

4. Make a neat copy of your project digraph, showing the critical tasks and earliest finish time. Also make a copy of the graph model for the committee scheduling problem, showing the graph coloring.

a. File these two graphs at the location in the classroom designated by your teacher. Examine the graphs filed by other groups and compare their graphs to those from your group.

b. Write a question to at least one group asking them to explain something about their work that you found interesting or did not understand. Answer any questions your group receives.

INVESTIGATION ▶3 Booths and Floor Plans

In this investigation, you will sketch a floor plan for the carnival and build a scale model of a booth.

1. Assume that the carnival will be held in a rectangular-shaped gym with dimensions 40 meters by 30 meters. Using centimeter graph paper, with 1 centimeter corresponding to 1 meter, sketch a floor plan for the carnival. Your floor plan should show the placement of the following items:

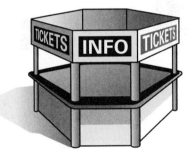

■ The ticket and information booth. This booth is hexagonal-shaped so that during peak times customers can line up at six windows to get information or tickets. Two sides of the booth have length 1 m and the other sides have length 2.24 m. The booth is to be placed in the center of the gym.

■ The game booths. There are ten game booths to be arranged along the sides of the gym. Eight of them are U-shaped, 1.5 m × 2 m × 1.5 m. The 2 m side faces out. The other two are triangular-shaped—two sides are the same length and the third side, which faces out, is 2 m long.

- Concessions. Six tables, each of which has a 2 m × 1 m table top, are placed in a U-shape to serve as the area where food is sold.

- Decide on one other feature of the floor plan. Describe it and add it to your plan.

2. Now consider the ticket booth in more detail.

   a. Build a model of the ticket booth frame using a scale of 10 cm to 1 m. The ticket booth has base dimensions as described in Activity 1. The walls of the booth are 2 m high.

   b. Design a tent-like canopy for the booth.

   c. Make a careful sketch of the ticket booth.

   d. How much canvas will be needed to cover the booth, including the canopy?

3. File your floor plan and your ticket booth model and sketch at the location in the classroom designated by your teacher. Study the floor plans, models, and sketches filed by other groups and compare these items to your group's work. Write a question to at least one group asking them to explain something about their work that you found interesting or did not understand. Answer any questions your group receives.

INVESTIGATION 4 ▶ Carnival T-Shirts

To help promote the carnival and raise more money, T-shirts will be designed and sold.

1. Design a logo for the carnival, to be used on the T-shirts as well as on other promotional materials. The logo might include the name of the school or a message of some sort. It could be an abstract design or a picture of something related to the carnival. It can be whimsical or serious. Be creative! The only requirement is that the logo must be symmetrical in some way.

   a. Describe the symmetry shown in your logo.

   b. Briefly explain why you chose your particular design and explain its meaning.

2. The price list for a T-shirt shop is shown below.

> ## Midwest Athletic Supply and Screen Printing
>
> **Set-up**
> $17.50 for one color $12.50 for each additional color
>
> **Art**
> $25 per hour
>
> **T-Shirts**
> Prices for T-Shirts with one color:
>
> | 1–15 | $6.55 ea. | 63–147 | $5.75 ea. |
> | 16–31 | $6.25 ea. | 148 or more | $5.40 ea. |
> | 32–62 | $5.95 ea. | | |
>
> Add $0.30 per shirt for each additional color

a. Suppose you decide to go with a deluxe four-color design that requires two hours of artwork from the shop's designers. Complete a table like the one below.

Cost of Four-Color T-Shirts

Number of T-Shirts Purchased	Total Cost
1	
10	
100	
200	
201	
202	

b. Suppose that you buy more than 150 T-shirts. Using *NOW* for the cost to buy a given number of shirts and *NEXT* for the cost to buy one more, write an equation showing the relationship between *NOW* and *NEXT*.

c. Using T for the number of T-shirts purchased and C for the total cost, write an equation showing the relationship between T and C for any number $T > 150$.

3. You plan to sell the T-shirts and make a profit. What price should you set for the shirts? Write a brief analysis justifying your choice of selling price. Your analysis should include the following:

■ An equation and graph showing the relationship between profit P and number of T-shirts sold T, where $T > 150$

■ An explanation of how to use the profit equation and graph to find the number of shirts you must sell to break even

■ An estimate of the profit you expect to make

■ A summary of why you chose your selling price

4. File a copy of your logo, an explanation of the symmetry it exhibits, and your profit analysis at the location in the classroom designated by your teacher. Examine the logos and solutions filed by other groups and compare these items to your group's work. Write a question to at least one group asking them to explain something about their work that you found interesting or did not understand. Answer any questions your group receives.

INVESTIGATION ▶ 5 Money Made and Money Spent

The main purpose of the carnival is to make money in a fun way. Both organizers and customers are concerned about the money aspect. The organizers want to know how much money the carnival will make. The customers want to know how much money they will spend.

1. Customers will buy tickets and then pay for the games using one or more tickets per game. Each ticket costs 25¢. This year, parents who buy tickets in advance can specify that the money paid goes to purchase equipment for their child's classroom. Parents are asking, "About how many tickets will I use at the carnival?"

At each of the last three carnivals, 25 parents were randomly chosen and asked how many tickets they used. These data are shown in the following table.

Number of Game Tickets Used by Sample of Parents at the Last Three Carnivals

3 Years Ago		2 Years Ago		Last Year	
26	20	20	32	6	8
20	30	48	24	4	10
20	18	36	12	4	12
16	25	16	11	20	10
4	8	16	36	8	18
21	15	44	40	8	20
20	14	24	8	8	12
20	48	40	12	24	28
26	16	8	42	7	26
10	10	8	12	12	24
16	16	18	22	22	20
20	35	10	14	18	12
34	—	44	—	16	—

a. Make a box plot of each year's data. Divide the work among your group. Draw all three box plots one below the other on the same axis.

b. Write a brief summary of the information shown in each box plot. Describe how patterns of ticket usage vary from year to year.

c. What other graphs or statistics might be useful to help you inform parents about how many tickets they will need?

d. When a parent asks you, "How many tickets will I use?" what will you say? Explain the reasoning behind your response.

2. Someone suggests the carnival can do more than support the school for the current year. The class agrees to set up a fund that will grow and support the school in future years. Your goal is to put $800 of the carnival profits into a savings account for the school. The account pays 5% interest compounded annually.

   a. How much money will be in the account when you graduate, assuming no withdrawals?

   b. Write equations that will allow you to calculate the balance of this account:

      ■ for any year, given the balance for the year before.

      ■ after any number of years x.

   c. Use the equations from Part b to make a table and a graph showing the growth of this account.

   d. Describe the pattern of growth in the savings account over a 10-year period.

   e. Think about how fast the account grows:

      ■ Is the account growing faster in year 5 or year 10? How can you tell?

      ■ What would the graph look like if the account were growing at a constant rate of change?

   f. Suppose that the school wants to buy a new computer for the library. A local manufacturer has offered to give the school a great price. The computer is expected to cost about $1,000. How long will it be before there is enough in the account to buy the computer?

3. Make a neat copy of your work on this investigation, including graphs, plots, tables, equations, and explanations. File the copy at the location in the classroom designated by your teacher. Check the solutions filed by other groups and compare them to your group's work. Write a question to at least one group asking them to explain something about their work that you found interesting or did not understand. Answer any questions your group receives.

INVESTIGATION ▶6 Ring-Toss Game

The most important part of a carnival is, of course, the games. Suppose that you are setting up a ring-toss game. A number of two-liter bottles of soda are lined up. The goal of the game is to toss a ring around the top of one of them. If a player hooks one of the bottles (which is called getting a "ringer"), then he or she gets to keep it. Your task is to design this game. You need to consider how large the rings should be, how far back the players should stand, and what the cost to play the game should be.

1. Examine the following data collected from a ring-tossing experiment. The data show the average number of ringers per 100 tosses at a distance of 1 meter from the bottles, for rings of varying diameters.

Sample Ring-Toss Data for 1-Meter Tosses

Diameter of Ring (in centimeters)	Average Number of Ringers per 100 Tosses
4	3
5	6
6	8
7	11
8	14
9	19
10	22
11	27
12	32

a. Make a scatterplot of the data.

b. Describe and explain any patterns or unusual features of the data.

c. Notice that when the ring diameter doubles, the number of ringers increases by a factor of about four. Can you suggest any explanation for this pattern in terms of the size of the circular rings?

2. Data collected from another ring-tossing experiment are shown in the following table. These data show the average number of ringers per 100 tosses with rings of diameter 12 cm, for players standing at varying distances from the bottles.

Sample Ring-Toss Data for 12-cm Diameter Rings

Distance from the Bottles (in meters)	Average Number of Ringers per 100 Tosses
1	32
2	15
3	8
4	4
5	1

a. Make a scatterplot of the data.

b. Describe any patterns you see in the data.

c. Suppose *NOW* is the average number of ringers at one of the distances in the table and *NEXT* is the average number of ringers for a distance 1 meter farther away. Write an equation that approximates the relationship between *NOW* and *NEXT*.

3. It is important to charge enough to play the game so that the income from ticket sales for the game is greater than the cost of the prizes given away. For this activity, assume that a ring with diameter 12 cm is used and players stand 2 meters away from the bottles.

a. Suppose that a local merchant offers to support the carnival by loaning you all the soda you need to set up your game. The merchant will charge you

60¢ for every bottle you give away as a prize. You decide to charge 25¢ for a toss. Based on the data in the tables, you expect 15% of the tosses to be ringers. Use T for the number of tosses by customers and P for your profit from the game. Write an equation that shows the relationship between P and T.

b. Given the arrangement with the local merchant in Part a, what is the least you can charge for a toss and still make a profit?

c. Suppose you have no sponsor and must pay the usual retail price for the soda. For stores near where you live, what is a reasonable price for a two-liter bottle of soda? In this situation, what is the least you can charge for a toss at a distance of 2 meters and still make a profit?

d. Based on the data in Activity 2, how would your profit-modeling equation in Part a change if players tossed rings at a distance of 3 meters? In this situation, what is the least you can charge for a toss and still make a profit?

e. As designers of fun, profitable games, would your group recommend a 2-meter toss or a 3-meter toss? Explain your reasoning.

4. Make a neat copy of your work on this investigation, showing graphs, equations, and other answers. File this "solution sheet" at the location in the classroom designated by your teacher. Study the solutions filed by other groups and compare them to your solutions. Write a question to at least one group asking them to explain something about their work that you found interesting or did not understand. Answer any questions your group receives.

INVESTIGATION ▶ 7 Free-Throw Game: Beat the Pro

Games of skill, especially those involving sports, are always popular at carnivals. Suppose that you are in charge of setting up a basketball free-throw game where a challenger pits his or her skill against a pro. In this case, the pro is the top free-throw shooter from the girls' basketball team. The challenger and the pro each shoot ten free throws. If the challenger makes more baskets than the pro, then he or she wins a prize. Your job is to decide how much to charge to play the game and what prizes should be awarded to winners.

1. You know from the basketball season's statistics that the pro makes about 85% of her free throws. What about the challenger's percentage of successful free throws? You cannot know the shooting percentage of every challenger that might play the game. However, it would be helpful to get some information to help you decide on the price to charge and prizes to award. One of your friends, who likes basketball and is a pretty good shooter, agrees to help you gather some data by being a sample challenger. He completes 50 trials of 10 free throws and records the number of baskets for each trial. These data are shown in the following table.

Number of Free Throws Made (Out of 10) for 50 Trials by One Sample Challenger

8	8	9	6	7	8	7	6	8	7
8	8	10	9	8	7	7	8	5	5
7	5	6	10	6	8	9	8	9	10
8	7	8	9	8	5	8	6	8	7
7	6	9	4	9	6	8	10	7	6

a. Construct a histogram for these data. Describe and explain any patterns in the distribution.

b. Based on the data, about what percent of free throws does the sample challenger make? Explain using a statistical measure.

c. Does he appear to be a fairly consistent shooter? Justify your answer by using an appropriate plot and summary statistics.

2. To help make good decisions about the amount you should charge and the value of prizes, you can simulate a game between the sample challenger and the pro.

a. What is a reasonable estimate for the probability that the pro will make a particular free throw?

b. Explain why 0.75 is a reasonable estimate for the probability that the sample challenger will make a particular free throw.

c. Using the probabilities above, design a simulation to estimate the probability that the challenger wins the game.

d. Conduct 5 trials of your simulation. Add your results to the Simulated Beat The Pro Game table so there is a total of 100 trials. (The table can be obtained from your teacher.)

3. Now examine the simulation table you completed in Activity 2.

  a. Construct a histogram of the difference between the number of free throws made by the challenger and the number made by the pro. Describe the shape of the distribution. Interpret the shape in terms of outcomes of the "Beat The Pro" game.

  b. What is your estimate of the probability that the challenger wins? Remember that the challenger wins only if he or she makes *more* of the 10 free throws than the pro does.

4. A local sporting goods store will support the carnival by selling you top-quality basketballs at a discount for prizes. The balls will cost $8 each.

  a. Based on the simulation data and your analysis in Activity 3, determine how much you should charge the sample challenger to play the game. Remember, you want to keep the game affordable and yet ensure a profit over the course of many games.

  b. The actual game will be played with many different challengers, not just the one sample challenger. What do you think is a good price to charge for the actual game?

5. Would the pro be more likely to beat a 75% free-throw shooter in a game with 20 shots or one with 10 shots? Explain your reasoning.

6. Make a neat copy of your work on this investigation. File it at the location in the classroom designated by your teacher. Examine the work filed by other groups in the class and compare it to your work. Write a question to at least one group asking them to explain something about their work that you found interesting or did not understand. Answer any questions your group receives.

INVESTIGATION ▶8▶ Further Analysis

In Investigations 2 through 7, you analyzed a variety of situations related to planning the carnival. Of course, there are other things to consider as well. Choose one of your ideas from Investigation 1 or from the "Think About This Situation" at the beginning of this Capstone. Carry out a brief mathematical analysis of the idea. Specifically, you should formulate and answer at least two questions related to your idea. For example, you might design and analyze another game, as is done in Investigations 6 and 7, or you might collect and analyze data on what kinds of games are most popular. File a copy of your analysis at the location designated by your teacher.

REPORTS: Putting It All Together

Finish this Capstone by preparing two reports, one oral group report and one individual written report as described below.

1. Your group should prepare a brief oral report on one of the investigations you have completed. You will present the report as if you are reporting to the principal of the elementary school that is planning to have the carnival. Your teacher will play the role of the principal. Your report should meet the following guidelines.

   ■ Choose one of the investigations you have completed. Confirm your choice with your teacher before beginning to prepare your report.

   ■ Examine the work that other groups have filed on your chosen investigation. Compare your work to theirs and discuss any differences with them. Modify your solutions, if you think you should.

   ■ Begin your presentation with a brief summary of your work on the investigation.

   ■ Convince the principal that your solutions are correct and should be adopted.

   ■ Be prepared to discuss alternative solutions, particularly those proposed by other groups that also worked on the same investigation.

   ■ Be prepared to answer any questions from the "principal" or your classmates.

2. On your own, write a two-page report summarizing how the mathematics you have learned in this course can be used to help plan a school carnival.

Checkpoint

In this course, you have learned important mathematical concepts and methods and you have gained valuable experience in thinking mathematically. Look back over the investigations you completed in this Capstone and consider some of the mathematical thinking you have done. For each of the following habits of mind, describe, if possible, an example where you found the habit to be helpful.

a Search for patterns

b Formulate or find a mathematical model

c Collect, analyze, and interpret data

d Make and check conjectures

e Describe and use an algorithm

f Visualize

g Simulate a situation

h Predict

i Experiment

j Make connections—between mathematics and the real world and within mathematics itself

k Use a variety of representations—like tables, graphs, equations, words, and physical models

Be prepared to share your examples and thinking with the entire class.

Index of Mathematical Topics

Index of Contexts

Photo Credits

We would like to thank the following for providing photographs of Core-Plus students in their schools. Many of these photographs appear throughout the text.

Janice Lee, Midland Valley High School, Langley, SC
Steve Matheos, Firestone High School, Akron, OH
Ann Post, Traverse City West Junior High School, Traverse City, MI
Alex Rachita, Ellet High School, Akron, OH
Judy Slezak, Prairie High School, Cedar Rapids, IA
The Core-Plus Mathematics Project

Cover, Photodisc: 325, Susan Van Etten; 326, John Colletti/Stock Boston; 330 (top left), Jeff Schultz/Alaska Stock Images, (center left) US Forest Service; 332, David Germen; 337, Tomas del Amo/West Stock; 340, Peter Menzel/Stock Boston; 343, Jack Demuth; 347, S. Wanke/West Stock; 348, Susan Van Etten; 350, AP/Wide World Photos; 353, (left) Art Walker/Chicago Tribune, (right) Bernard Gotfyd/Woodfin Camp & Associates; 355, (left) Chicago Tribune, (right) Angelo Hornak/CORBIS; 363, Susan Van Etten; 367, Chicago Tribune; 368, Thomas Schmitt/Image Bank; 369, Jack Demuth; 372, Carl Hugare/Chicago Tribune; 373, (right) Joe Branneis; 380, Quentin Dodt/Chicago Tribune; 381, Karen Engstrom/Chicago Tribune; 385, Jose Moré/Chicago Tribune; 393, 394, R. Buckminster Fuller; 395, Susan Van Etten; 396, Manfred Kage/Peter Arnold, Inc.; 401, "Kualoa" Helen M. Friend, 1991; 402, California Academy of Sciences, the Elkus Collection (#370-1820) Maricopa Black-on-Red Jar by Mary Juan, ca. 1940's, Southern Arizona; 404, AP/Wide World Photos; 405 (top), Alinari/Art Resource, NY; 409, (top) © The British Museum, Department of Ethnography, Museum of Mankind; (bottom) © Dover Publications; 416, UNI-GROUP U.S.A.; 419, Carlyn Iverson; 422, Photo courtesy Motorola; 424, Jack Demuth; 425, Texas Instruments Incorporated, Dallas, Texas; 428, Manfred Kage/Peter Arnold, Inc.; 431, Akos Szilvasi/Stock Boston; 433, AP/Newsfeatures Photo; 434, Texas Instruments Incorporated, Dallas, Texas; 435, Blair Seitz/West Stock; 439, AP/Wide World Photos; 440, Henry Groskinsky/Peter Arnold, Inc.; 448, Chuck Berman/Chicago Tribune; 450, John Colletti/Stock Boston; 453, Chicago Tribune; 454, Jim Brown; 456, Chip Henderson/Tony Stone Images; 457, UPI/Bettmann/ CORBIS; 461, M.&C. Ederegger/ Perer Arnold, Inc.; 463, Jack Demuth; 467, AP/Wide World Photos; 468, 469, Jack Demuth; 470, Kyle Summers/Washington and Jane Smith Home; 471, Dianne Bell/Peter Arnold Inc.; 473, Don Casper/Chicago Tribune; 474, Michael Collier/Stock Boston; 475, Brent Jones/Chicago Tribune; 478, 483 AP/Wide World Photos; 484, Ron Kotulak/Chicago Tribune; 485, Cindy and Grace Chang; 490, Carl Wagner/Chicago Tribune; 492 (left), AP/Wide World Photos, (right) Matt Meadows; 493, PhotoDisc; 494, Richard Clintsman/ Tony Stone Images; 495, 501, Jack Demuth; 502, Texas Instruments Incorporated, Dallas, Texas; 503, 504 (left), Jack Demuth; 504 (right), James K. Laser; 505, Chris Walker/Chicago Tribune; 508, "Used with permission from McDonald's Corporation."; 511 (left), Chicago Tribune, (right) Matt Meadows; 512, DRABBLE reprinted by permission of United Feature Syndicate, Inc.; 513, AP/Wide World Photos; 515, Lori Adamski Peek/Tony Stone Images; 516, Reuters New Media Inc./CORBIS; 518, AP/Wide World Photos; 519, EyeWire/Getty Images; 521, Stone/Getty Images; 524, Bettmann/CORBIS; 525 (left), Ernie Cox Jr./Chicago Tribune, (right) UPI/Corbis-Bettmann; 526, Billie E. Barnes/Stock Boston; 527, (top) Susan Van Etten; (bottom) Chris Walker/*Chicago Tribune*; 530, Christine Longcore; 532, Billy E. Barnes/Stock Boston; 534, Jack Demuth; 535, John Irvine/Chicago Tribune; 537, Phil Greer/Chicago Tribune; 541, Susan Van Etten; 542, Texas Instruments Incorporated, Dallas, Texas; 545, Heidi Kolk